M.A.Buth

3D Drucker: Patente & Erfindungen

M.A.Buth

3D Drucker: Patente & Erfindungen

1. Auflage

Stand: 24.09.13

Hinweis zur Erstauflage:
Trotz aller gebotener Sorgfalt, kann bei einer
Erstausgabe nie ausgeschlossen werden, dass Fehler
übersehen wurden. Wir haben die Aktualität des
Themas gegen dieses Risiko abgewogen und bieten
jedem Käufer der Erstauflage, gegen Einsendung des
Buches, einen kostenlose Ersatz (Update) der
Folgeauflage an.
Einsendungen bitte frei (Büchersendung 1€), unter
Angabe der Empfängeradresse, an den Verlag

Verlag

ViSdP
adrenalinemedia
vertreten durch
Marcel A. Buth
Neue Strasse 32b
61250 Usingen
adrenalinemedia@web.de

Inhaltsverzeichnis

Vorwort

Anfang der 1970er Jahre tauchen die ersten „echten" 3D Drucker Patente auf. Vorläufer gab es schon in den 50er Jahren. Seither haben sich eine Vielzahl von Erfindungen damit beschäftigt, wie man Dinge erschafft. Das Patentwesen ist jedoch ein Datenlabyrinth mit babylonischer Sprachenvielfalt und biblischem Ausmaß. Entgegen unseren heutigen Gewohnheiten sind Patente nicht rationell per Suchmaschine recherchierbar. Zu viele unterschiedliche Datenbanken aus zahlreichen Nationen benutzen unterschiedlichste Formate und Zugänge. Altertümliche IT Systeme mit kryptischen Suchfunktionen lassen den interessierten Leser verzweifeln. Hinter englischen nichtssagenden Titeln verbergen sich jedoch oftmals wahre Schätze an Informationen, detaillierte technische Zeichnungen, chemische Formeln und vieles mehr.

Patentrecherche ist und bleibt daher Handarbeit. Ein mühseliges, manuelles Durchwühlen von Datenbanken auf allen Kontinenten und Sprachen. Diese Vorarbeit leistet dieses Buch. Es ist Startpunkt für eigene Nachforschungen. Indem es interessante und wichtige Patente nennt, da wo erforderlich in einer kurzen Übersetzung zusammenfasst und ergänzende Informationen gibt.

Viele Patente sind heute abgelaufen oder verwaist, weil Firmen längst untergegangen oder Gebühren nicht bezahlt wurden. Sie sind also für eigene – auch gewerbliche - Nutzungen frei!

Patente sind keine Geheimdokumente, auch wenn Unternehmen dies gerne so hätten. Hier müssen sie die Karten auf dem Tisch legen und erklären wie ihre Produkte funktionieren. Es sind öffentliche Dokumente, die jedem Bürger frei zur Einsichtnahme stehen.

Er muss sie nur nutzen.

Marcel A. Buth   Bad Homburg den 14.09.2013

6 Patente und Schutzrechte

Patente und Schutzrechte

Darf man Patente nachbauen?

Diese Frage kann man indirekt mit „Ja" beantworten. Patente schützen den Inhabern nicht vor dem Nachbauen an sich, sondern nur vor einem finanziellen Schaden durch entgangene Gewinne. Dies wird aber bei einer rein privaten Nutzung nicht angenommen. Wer also für eigene Zwecke die Informationen patentierter Verfahren nutzt, braucht sich keine weiteren Gedanken zu machen. Wer aber patentierte Produkte nachbaut und verkauft, schadet dem Inhaber des Schutzrechtes, denn ihm entgehen so Einnahmen und die kann er als Vermögensschaden einfordern.

Trotzdem ist es auch für Hersteller interessant, die Patente der Konkurrenz zu verfolgen. Vielleicht geben sie ja wertvolle Hilfen beim Entwickeln eigener Ideen? Zudem laufen Patente irgendwann einmal aus und Nachbauten wären dann legal möglich.

Wer also für sich, die Familie oder einen guten Freund, unentgeltlich ein patentiertes Verfahren oder Gerät nachbildet, braucht sich keine Sorgen zu machen. Vielleicht fällt ihm dabei sogar noch eine Verbesserung oder Abwandlung ein, die dann eine völlig eigenständige Lösung darstellt und sich dann sogar wieder patentieren lässt!

Mit dem Teilen von Plänen von Produkten auf Plattformen wie beispielsweise Thingiverse, sollte man jedoch vorsichtig sein, da man dort eventuell einen Rechteinhaber schädigt, selbst wenn einem das nicht bewusst ist. Daher sollte man immer zu überprüfen, ob die geniale Idee die man hatte, nicht doch schon von einem anderen hellen Kopf ersonnen wurde, denn auch bei Patentverletzungen gilt:

Unwissenheit schützt vor Strafe nicht!

8 Patente und Recherche

Patente und Recherche

Dem Eifer des hochmotivierten Unternehmensgründer stehen eventuell bestehende Schutzrechte Dritter gegenüber. Das ist ein weiterer Grund, weshalb FDM 3D Drucker für kleine Projekte die bessere Wahl sind: Hier wurde bereits vieles veröffentlicht was fortan nicht mehr schutzwürdig und patentierbar ist.

Führende Unternehmen haben sich ihre teure Entwicklung natürlich mit entsprechenden Patenten abgesichert um zu verhindern, dass Nachahmer von ihren Erkenntnissen profitieren. Daher gilt es insbesondere vor dem Veröffentlichen von Produktspezifikationen, oder gar dem Anbieten von Produkten, die entsprechenden Patentschriften sorgfältig zu studieren. Was zunächst wie eine mühsame Bürokratie a lá Buchhaltung erscheint, erweist sich zum Quell guter Ideen und nützlicher Informationen. Patentinhaber müssen nämlich in ihren Schriften Verweise – sogenannte Patentzitate - zu bestehenden Patenten und Schutzrechten, sowie den allgemeinen Stand der Technik angeben, um sich mit ihrer Erfindung deutlich von diesen abgrenzen zu können. Also führt das Studium der Patente zu weiteren Erkenntnissen sowie interessanten Informationen und **sollte vom Unternehmer ausgiebig genutzt werden**!

Die Tatsache allein, dass ein Patent auf eine bestimmte Technologie oder Erfindung besteht, bedeutet nicht automatisch dass man diese nicht nutzen kann. Hier gibt es viele Ausnahmen und Lücken. Da dieses Buch keine juristische Patentberatung bieten kann – hierzu erfordert es immer eine Konsultation mit einem entsprechendem Fachanwalt - können die Möglichkeiten nur angedeutet werden, die dem Tatendrang des Startups entgegen stehen.

Zunächst einmal wird ein Patent, sofern es formal richtig eingereicht und überprüft wurde, vom Patentamt angenommen. Sollte ein anderer Rechteinhaber der Meinung sein, dass er eine Erfindung zuerst gemacht hat, so **kann ein Patent auch angefochten werden**. Diesen kostspieligen, langwierigen und unsicheren Prozessweg sollte man aber den großen Unternehmen oder Open Source Organisationen überlassen.

Weiter gibt es Zeitgenossen, die scheinbar selbstverständliche Vorrichtungen patentieren lassen, nur in der Hoffnung, dass man hiermit eines Tages andere Unternehmen erpressen kann, die diese Lösungen nutzen. Man spricht hier von sogenannten **Trivialpatenten**. Zu finden sind sie (leider) bevorzugt im IT Bereich. Die "Geschäftsidee" solcher sogenannter **Patent Trolle** entspringt wohl dem Geist des "Domaingrabbings" und anderer Unsitten aus der Frühzeit des Internets. Die Idee ähnelt einem Investor, der vorsorglich sein Claim absteckt, da es ja irgendwann einmal sein könnte, dass man dort Gold oder Öl findet, ohne jedoch die Absicht zu haben die Rechte tatsächlich zu nutzen. Ein derartiges Verhalten schadet allerdings dem globalen technischen Fortschritt und somit der gesamten Menschheit. Rechtlich gesehen bieten diese Patente einen Angriffspunkt, hinsichtlich der **Erfindungshöhe**. Diese muss – neben der Neuheit der Idee - gegeben sein um eine Erfindung patentwürdig zu machen. Die Erfindungshöhe ist ein recht abstrakter Begriff, der besagt, dass eine Erfindung überdurchschnittlich innovativ sein muss, nicht also lediglich die Wiedergabe bereits allgemein bekannter Zusammenhänge. Eine Erfindung besitzt dann eine gewisse "Höhe" wenn sie nicht ohne weiteres von einem anderen Fachmann ebenso hätte erdacht werden können. Solche Patente werden individuell, aber auch durch Initiativen in der EU und den USA von Organisationen angegriffen. Gerichtsentscheidung zu oder besser gegen solche Trivialpatente finden sich mittlerweile und es besteht Hoffnung, dass in Zukunft das Geschäft mit der Patentabzocke ein Ende nimmt.

Manche Patentrechte werden aber auch vom Inhaber gar nicht wahrgenommen. In der Softwarebranche gab es hier großzügige Regelungen von Großunternehmen um mit dem boomenden Open Source und Linux Markt weiter wachsen zu können. Dort sind Patente äußerst unbeliebt, da sie die freie Entwicklung hemmen und umgangen werden müssen. Hier gilt es sich zu erkundigen, vielleicht hat der Patentinhaber keine Verwertungsinteressen mehr und hat dies auch bekannt gegeben. Vorsicht aber bei direktem Kontakt zum Rechteinhaber! Eine solche Anfrage landet mit Sicherheit in der Rechtsabteilung des Unternehmens und ein unterbeschäftigter Anwalt findet sich bestimmt immer, der sich mal profilieren möchte. Ist man erst mal auf dem Radar solcher Abteilungen oder Kanzleien, wird man auf Schritt und Tritt beobachtet...

Es gibt Fristen innerhalb derer sich ein Rechteinhaber um die Wahrnehmung seiner Patentrechte kümmern muss. Erfährt er von einem Verstoß, so hat er maximal 2 Jahre Zeit diesem nachzugehen und seine Rechte durchzusetzen. Tut er dies nicht, riskiert er den Verfall seiner Ansprüche. Ein Grund mehr, das kleinere Unternehmen es sich sehr genau überlegen sollten ob sie ihre Wettbewerbsvorteile nur auf Patenten aufbauen. Dies erfordert nämlich immer zusätzlich den kostspieligen Einsatz einer Patentüberwachungskanzlei, die weltweit nach Verstößen fahndet und diese verfolgt.

Der Autor selbst hat mehrfach in seiner Zeit als Elektronikentwickler und Produzent, Schreiben von Anwälten aus den USA und auch Deutschland erhalten, in denen Patentansprüche angemeldet und mittels Patentschrift begründet wurden. Dies nicht, weil er Produkte kopiert oder nachgemacht hatte, sondern einfach weil entweder bei gleichen Problemstellungen oftmals ähnliche Lösungen heraus kommen, die Patentrecherche nicht oder nur unzureichend durchgeführt wurde, oder weil ein Kaufteil eines Vorlieferanten offensichtlich ein Plagiat waren. Wenngleich die telefonbuchschweren Schreiben von Anwaltskanzleien aus New York die Stimmung des Arbeitstages deutlich trübten, gab es in all den Jahren nicht einen einzigen Fall, in dem nach dem ersten Schreiben weitere Schritte folgten. Dies liegt sicherlich auch darin begründet, dass in der kurzlebigen Technikwelt von heute, die Verfolgung von Patentansprüchen nur in den seltensten Fällen wirtschaftlich lukrativ für den Rechteinhaber sind. Bis ein Gerichtsurteil vorliegt sind Jahre vergangen, das Produkt ist veraltet und unverkäuflich und eine Menge Kosten sind für den Rechteinhaber angefallen. Freuen können sich hierüber höchstens die beauftragten Anwälte.

Weiterhin lassen sich **Patente** oftmals auch **umgehen**. Hierzu müssen die vom Rechteinhaber geäußerten Ansprüche sorgfältig studiert werden. In den **Ansprüchen** formuliert der Antragsteller, welche Vorrichtungen oder Besonderheiten an seiner Lösung explizit geschützt werden sollen.

Darüber hinaus werden Patente immer nur für bestimmte nationale Zonen erteilt. Ein Patent aus den USA, das nicht für Deutschland beantragt wurde, ist somit in Deutschland auch nicht gültig und umgekehrt. Zu den hohen Kosten für eine ausländische Patentschrift, kommen noch erhebliche Gebühren für die Übersetzung und jährliche Aufrechterhaltungskosten, die mit der Zeit steigen. Daher machen sich viele Patentinhaber nicht die Mühe und die Kosten, ihre Erfindung auf sämtlichen Kontinenten der Welt anzumelden.

Eine intensive Recherche von ausländischen Patentschriften kann also erhebliche Vorteile bringen. Übrigens kann eine solche Erfindung, die nur in einem Land angemeldet wurde, dann auch nicht mehr im eigenen oder einem fremden Land angemeldet werden, da durch die Veröffentlichung – egal wo - ein wesentliches Merkmal der

Patentwürdigkeit verloren geht: Die Neuheit.

Erste Patentstreitigkeiten gibt es bereits zwischen der Makergemeinde und der realen Wirtschaft. So hat die Pionierfirma 3D Systems, die den ersten Stereolithographiedrucker produzierte, die Crowdfunding Plattform Kickstarter und die Betreiber des Projekts Formlabs wegen Patentverstößen verklagt.

"3D Systems ist Pionier und Erfinder des 3D Drucks mittels Stereolithographie und hält viele gültige Patente, die verschiedene Merkmale des Stereolithographie Prozesses abdecken.(...) Obwohl Formlabs öffentlich erklärt hat, dass einige Patente abgelaufen seien, glaubt 3D Systems, dass der Form1 3D Drucker mindestens eines unserer Patente verletzt und wir beabsichtigen, unsere Patentrechte durchzusetzen." so Andrew Johnson, General Counsel von 3D Systems.

Es ist also Vorsicht geboten, wenn man im öffentlichen Raum, Produkte anbietet die einem Patentschutz unterliegen könnten!

Wer sicher gehen will, muss vor dem Veröffentlichen von Bauteilen oder gar ganzen Produkten, zunächst einmal die Patentlage klären. Hierzu muss im jeweiligen nationalen Rechtsgebiet, in dem die Erzeugnisse angeboten werden, in den Patentschriften recherchiert werden. Das ist nicht immer ganz einfach, da es sich um winzige Details handeln kann, die ein Anmelder beansprucht. Auch können Patente wirksam sein, die aus ganz anderen technischen Bereichen stammen, wie z.B. dem Werkzeugmaschinenbau, der Feinwerktechnik der Kunststofftechnik usw.

12 Patente und Recherche

Patentrecherche

Folgende Datenbanken bieten online die Möglichkeiten zur Recherche:

USA:

http://www.uspto.gov/patents/process/search/index.jsp

Europa und Deutschland:

http://www.dpma.de/patent/recherche

Hat man einmal die Anmeldenummer des Patents recherchiert, so kann man beispielsweise über Google unter dem Suchstring „Patent + *Anmeldenummer*" eine sehr gute Darstellung der Patentschriften erhalten, die derjenigen der Patentämter sogar überlegen ist. Beim US Patentamt benötigt man z.B. ein Quicktime Plugin um Zeichnungen zur Anmeldung sehen zu können. Bei Google sind die Abbildungen im PNG Format schnell und in jedem Browser ladbar. Allerdings sind die Metadaten bei Google stark ausgedünnt und die Verweise zu anderen Patenten und Nichtpatenten nicht verlinkt. Daher eignet sich diese Darstellung nur, wenn man schon genau weiß welches Patent man sucht. Schön hingegen, dass Google auf einer Seite sämtliche Übersetzungen verlinkt hat. Man kann also z.B. von der Original US Anmeldenummer einfach auf die deutsche Übersetzung umschalten, sofern das Patent auch in Deutschland angemeldet wurde.

Es gibt weitere gute Dienste die Patentschriften auswerten. Da wäre beispielsweise http://www.patentbuddy.com . Hier erhält man nicht nur optisch gut aufbereitete Patentschriften, es werden auch umfangreiche Statistiken dargestellt, die sich nicht direkt aus dem Patent ableiten und somit einen Mehrwert bieten. Man erfährt so, wie oft das aufgerufene Patent von anderen Patentschriften zitiert wird, wie viele weitere Patente die Anmelder noch halten und welche und einiges mehr.

Interessant kann auch die exakte Angabe der jährlichen Kosten sein, die das Patent dem Inhaber verursacht und ob er die Rechnungen beglichen hat. Hier kann man auch erkennen, ob er es tatsächlich noch ein Interesse an der Verfolgung seiner Patentrechte hat.

Patentschriften sind wahren Fundgruben für den technisch interessierten Leser. Nicht nur, dass man durch sie über Verfahren und Lösungen erfährt, von denen man in der Presse oder der Öffentlichkeit noch nie gehört hat -In jeder Patentanmeldung steckt auch ein ungeheurer Rechercheaufwand, den man für sich selbst nutzen kann. So verweisen Anmelder – aus eigenem Interesse – nämlich auf möglichst alle Patentanmeldungen die mit der Erfindung zu zu haben. Sie erwähnen diese weil sie sich deutlich von deren Erfindungen abgrenzen und und somit einer negativen Patentprüfung entgehen wollen. Befinden die Prüfer nämlich, dass die hier vorliegende Patentanmeldung ganz oder in Teilen bereits von anderen Patenten beansprucht wird, fehlt die Neuheit der Erfindung und die Erteilung des Patents wird abgelehnt.

Oftmals finden sich auch Angaben zu Buchtiteln oder wissenschaftlichen Arbeiten zu dem Thema. Hier müssen Autor, Erscheinungsform usw. angegeben werden, so dass der Prüfer im Zweifelsfall diese Behauptungen überprüfen kann. Die Patentrecherche ist und bleibt die aufwändigste Arbeit im gesamten Patentverfahren, wenn man sicher gehen will, dass es die eigene Erfindung nicht doch schon gegeben hat. Daher sollte man sich diese Menge Arbeit, die sich ein anderer gemacht hat zunutze machen und seinen Spuren folgen um das eigene Wissen auf dem Gebiet zu vertiefen.

Eine Patentanmeldung verzweigt also durch ihre Bezugnahme auf andere Patente, Bücher oder wissenschaftliche Aufsätze ähnlich wie das Internet mit seinen Hyperlinks in einen ganzen Kosmos von Informationen. **Nutze ihn!**

Wurde ein Patent beantragt, bislang aber noch nicht erteilt, ist dies aus der Angabe Publikationstyp ersichtlich.

Auch ein Blick auf den Gebührenstatus lohnt: Wurden Patentgebühren nicht mehr gezahlt ist dies im Feld Gebührenstatus ersichtlich. Werden Gebühren nicht bezahlt, kann das Schutzrecht des Patent verfallen und auch nie wieder angemeldet werden. Somit gibt es niemanden der genau dieses Patent für sich beanspruchen kann. Über den genauen Status sollte man sich jedoch beim Patentamt erkundigen, da es auch Ausnahmen und Härtefälle geben kann.

14 Patente und Recherche

Systematische Suche

Wer in Zeiten von Google glaubt, er könne Wissen zu Patenten einfach durch die Eingabe von „Patent" + „3D" in der Suchmaschine erlangen, der irrt.

Die Patentmeldungen der USA/Canada und der EU allein, sind ein Ozean an Informationen. Dort einen einzelnen Hering zu finden ist eine äußerst zeitaufwändige, wenn nicht sogar unmögliche Methode.

Der einzige systematische Weg ist die hierarchische Suche von oben nach unten (Top-Bottom Prinzip). Man wählt zunächst den bekanntesten Marktführer zu dem Thema von Interesse. Diese Firmen haben viel Kapital und geben auch viel Geld für Patente aus. Wenn sie dies tun, dann haben Heerscharen von Patentanwälten und Ingenieuren bereits umfangreichste Recherchen durchgeführt die über reine Googleanfragen weit hinaus gehen. Ihnen stehen sämtliche Fachzeitschriften, wissenschaftliche Aufsätze, Bücher etc. zum Thema zur Verfügung. Das sind Informationsquellen, die man über eine reine Internetrecherche nie vollständig finden wird. Was liegt also näher, als diese Leute für sich arbeiten zu lassen?

In den Zitaten eines Patents finden sich die wichtigen Verweise auf andere Patente oder Fachliteratur etc. So gelangt man vom Marktführer über dessen bereits geleistet Recherchearbeit zu anderen relevanten Informationen ohne Streu- und vor allem Zeitverluste.

Trifft man im Rahmen der Recherche immer häufiger auf bekannte Informationsquellen oder Patente, hat man sich im Kreis gedreht und das Thema hinreichend sicher abgedeckt.

Ein weiterer Grund hierarchisch und nicht randomisiert vorzugehen, ist der Umstand dass kaum eine Erfindung aus dem Nichts ersteht und manche Teilaspekte oftmals schon in anderen Technologien lange bekannt sind. Dort eventuell aber unter einem ganz anderen Namen und somit für den Suchstring unauffindbar. Die Datenbanken des deutschen Patentamts DEPATISnet usw. bilden da keine Ausnahme.

Nur wenn man bereits weiß, was man sucht, wird man finden was man eigentlich schon weiß.

Strategische Patentrecherche

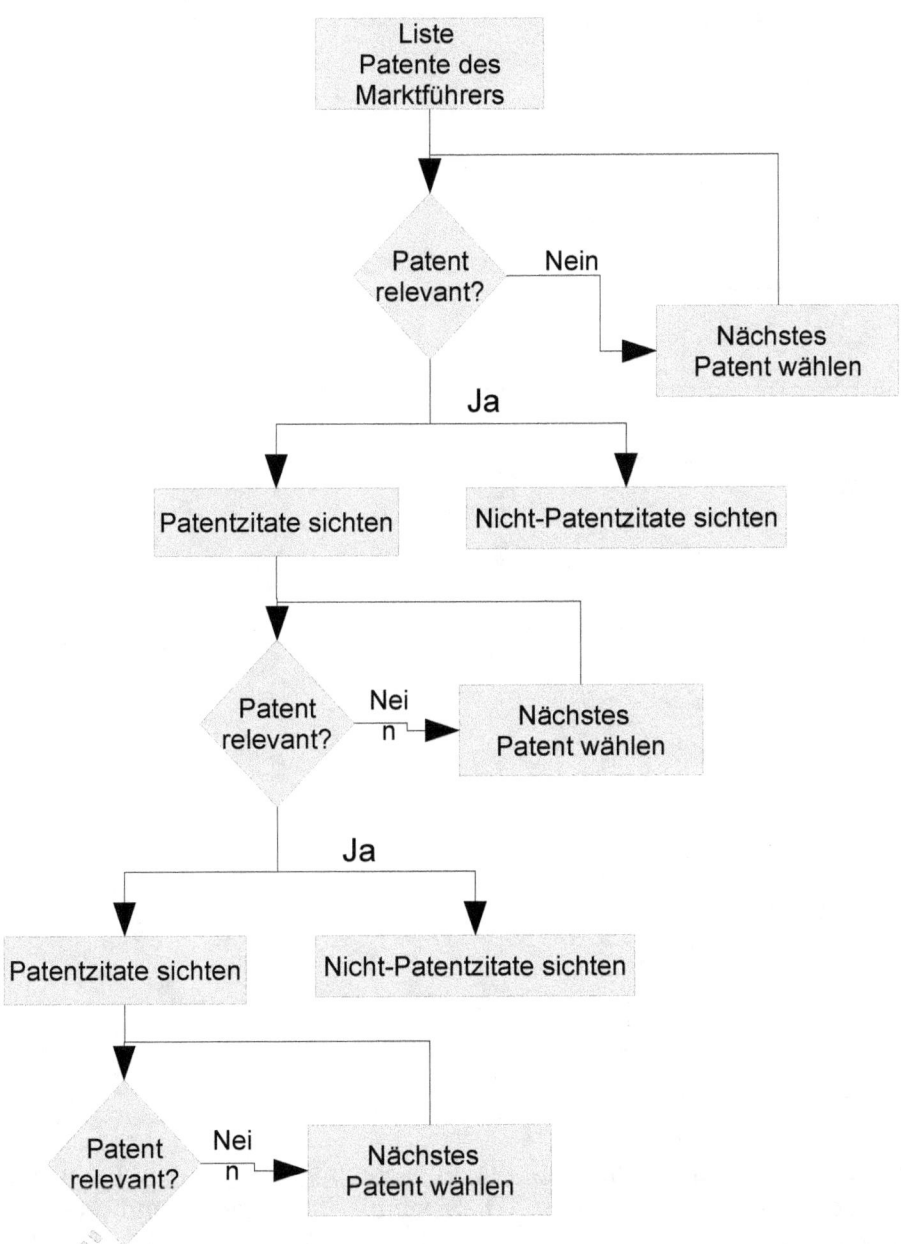

adrenalinemedia | Erstellt: M.A. Buth | Datum: 31.07.2013

16 Patente und Recherche

Pionier, Marktführer und auch noch Patentspitzenreiter ist die Firma 3D System. Sie hält die meisten Patente auf dem Gebiet der 3D Drucker nämlich 450. **70 dieser Patente sind bis dato abgelaufen** und lohnen eine genauere Inspektion.

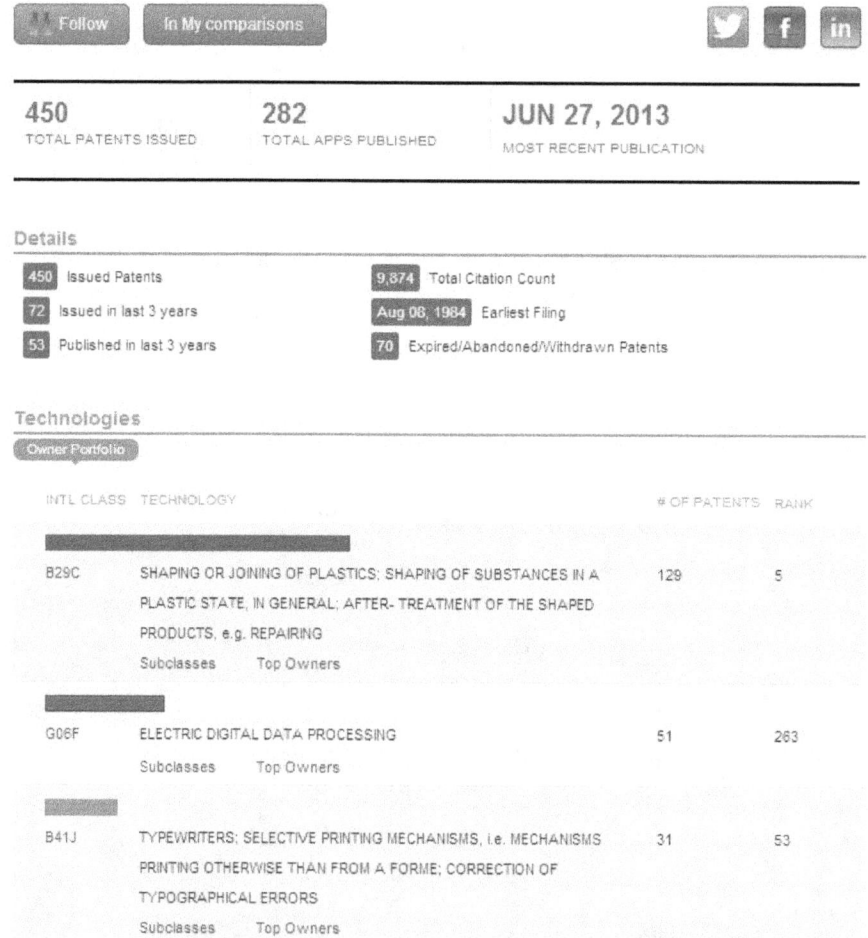

3D SYSTEMS, INC.
Patent Owner

Follow   In My comparisons

450	282	JUN 27, 2013
TOTAL PATENTS ISSUED	TOTAL APPS PUBLISHED	MOST RECENT PUBLICATION

Details

450	Issued Patents	9,874	Total Citation Count
72	Issued in last 3 years	Aug 06, 1984	Earliest Filing
53	Published in last 3 years	70	Expired/Abandoned/Withdrawn Patents

Technologies

Owner Portfolio

INTL CLASS	TECHNOLOGY	# OF PATENTS	RANK
B29C	SHAPING OR JOINING OF PLASTICS; SHAPING OF SUBSTANCES IN A PLASTIC STATE, IN GENERAL; AFTER- TREATMENT OF THE SHAPED PRODUCTS, e.g. REPAIRING	129	5
	Subclasses Top Owners		
G06F	ELECTRIC DIGITAL DATA PROCESSING	51	263
	Subclasses Top Owners		
B41J	TYPEWRITERS; SELECTIVE PRINTING MECHANISMS, i.e. MECHANISMS PRINTING OTHERWISE THAN FROM A FORME; CORRECTION OF TYPOGRAPHICAL ERRORS	31	53
	Subclasses Top Owners		

Erstaunlicherweise behandeln viele der Patente, längst nicht mehr nur die Stereolithographie, für die 3D Systems ja bekannt und erfolgreich geworden ist. Eigentlich meldet die Firma zu so ziemlich jedem Verfahren am Markt Patente, also auch dem Inkjetdrucken beispielsweise. Aber es ist eine Strategie heutiger Techfirmen, möglichst viele Patente anzusammeln, für die „Zeit danach". Dann nämlich wenn die Gewinne nicht mehr so üppig sprudeln, da die Konkurrenz an der Marktpostion nagt. Dann lässt sich noch Jahrzehnte nach dem Ableben des eigentlichen Geschäfts noch viel Geld mit Lizenzen oder Schadenersatzklagen machen.

STRATASYS, INC.
Patent Owner

Follow	In My Comparisons	① 2 Status Updates

114	103	JUL 16, 2013
TOTAL PATENTS ISSUED	TOTAL APPS PUBLISHED	MOST RECENT PUBLICATION

Details

114 Issued Patents	2,678 Total Citation Count
44 Issued in last 3 years	Oct 21, 1986 Earliest Filing
48 Published in last 3 years	2 Expired/Abandoned/Withdrawn Patents

Technologies

Owner Portfolio

INTL CLASS	TECHNOLOGY	# OF PATENTS	RANK
B29C	SHAPING OR JOINING OF PLASTICS; SHAPING OF SUBSTANCES IN A PLASTIC STATE, IN GENERAL; AFTER- TREATMENT OF THE SHAPED PRODUCTS, e.g. REPAIRING Subclasses Top Owners	44	32
G06F	ELECTRIC DIGITAL DATA PROCESSING Subclasses Top Owners	19	295
B28B	SHAPING CLAY OR OTHER CERAMIC COMPOSITIONS, SLAG OR MIXTURES CONTAINING CEMENTITIOUS MATERIAL, e.g. PLASTER Subclasses Top Owners	9	10

18 Patente und Recherche

Oder man verkauft die Patente an die aufstrebenden Wettbewerber. In den Zeiten des echten Goldrausches nannte man so etwas „Claim abstecken".

Auf Platz 2 folgt Stratasys mit weitem Abstand mit 114 Patenten. Die Anmeldung und der Unterhalt von Patenten ist eine teure Angelegenheit und so spiegelt die Anzahl der Patente im Bereich 3D Drucker auch die Finanzkraft und Manpower wieder, die die Unternehmen haben.

Technologies

Owner Portfolio   Top Owners

◄ back

		# OF PATENTS	RANK
B29C	SHAPING OR JOINING OF PLASTICS; SHAPING OF SUBSTANCES IN A PLASTIC STATE, IN GENERAL; AFTER- TREATMENT OF THE SHAPED PRODUCTS, e.g. REPAIRING		
	HUSKY INJECTION MOLDING SYSTEMS LTD.	277	1
	MOLD-MASTERS (2007) LIMITED	255	2
	3M INNOVATIVE PROPERTIES COMPANY	174	3
	THE BOEING COMPANY	160	4
	3D SYSTEMS, INC.	129	5
	E. I. DU PONT DE NEMOURS AND COMPANY	125	6
	FUJIFILM CORPORATION	124	7
	THE PROCTER & GAMBLE COMPANY	97	8
	NISSEI PLASTIC INDUSTRIAL CO., LTD.	96	9
	HON HAI PRECISION INDUSTRY CO., LTD.	93	10

Makerbot, erst 2006 aus dem Open Hardware Source Projekt RepRaP entstanden, verfügt immerhin schon über 6 Patente und hat 17 weitere angemeldet.Ansonsten gibt es noch etwas „Streubesitz" bei Firmen wie IBM oder Sony, bei denen man aber nicht weiß ob sie tatsächlich einmal in das Geschäft einsteigen werden, da oder einfach eine Firma mitsamt Schutzrechten übernehmen.In der Kategorie B29C, die für das Formen oder Fügen von Kunststoffen steht hat 3D System einen respektablen 5. Platz erklommen, direkt hinter Boing und noch vor dem Plastik und Sprengstoffmagnat Du Pont de Nemours. Das lässt ein wenig vermuten, was 3D Systems für die Zukunft so vor hat in dem Bereich und dass die Pläne wohl über das reine Brutzeln von Harz mit einem Laser hinaus gehen.

Gebrauchsmusterschutz

Das Gebrauchsmuster ist der "kleine Bruder" des Patents. Die Anmeldung ist stark vereinfacht, die Kosten sind geringer, allerdings ist auch der Schutz leichter zu umgehen.

Die Schutzvoraussetzungen für das Gebrauchsmuster sind denen für das Patent ähnlich. Durch ein Gebrauchsmuster können in Deutschland und Österreich gewerblich anwendbare Erfindungen geschützt werden, die neu sind und auf einem erfinderischen Schritt beruhen (DE: § 1 Abs. 1 GebrMG; AT: § 1 Abs. 1 GMG). Die Schweiz kennt keinen Gebrauchsmusterschutz. In den übrigen europäischen Staaten sind dem Gebrauchsmuster entsprechende Institute vor allem in Spanien von Bedeutung, aber in zahlreichen anderen Ländern vorgesehen, zum Teil mit Anforderungen wie beim Patent (Frankreich, Belgien, Ungarn), zum Teil mehr oder weniger stark abweichend.

Bei der Anforderung der „Neuheit" zeigen sich Unterschiede zum Patent:

Eine Erfindung ist neu im Sinne des GebrMG, wenn sie – zum Zeitpunkt der Anmeldung des Gebrauchsmusters – aus dem Stand der Technik noch nicht bekannt ist. Im Gegensatz zum PatG in diesem Sinne ist jedoch nur das bekannt, was schriftlich vorbeschrieben ist oder bereits im Inland vorbenutzt wurde (DE: § 3 Abs. 1 GebrMG; abweichend AT: § 3 Abs. 1 GMG: auch mündliche Beschreibungen, keine Beschränkung auf das Inland).

Darüber hinaus bleiben auch Veröffentlichungen bei der Prüfung der Neuheit unberücksichtigt, die durch den Erfinder oder seinen Rechtsnachfolger bis zu 6 Monaten vor der Anmeldung erfolgt sind (Neuheitsschonfrist; DE: § 3 Abs. 1 Satz 3 GebrMG; AT: § 3 Abs. 4 Nr. 1 GMG). Außerdem kann in Deutschland für eine Anmeldung innerhalb von sechs Monaten nach einer Ausstellung auf einer anerkannten Messe (im Bundesgesetzblatt veröffentlicht) eine „Ausstellungspriorität" in Anspruch genommen werden, so dass bei der Beurteilung der Schutzfähigkeit des Gebrauchsmusters alle Veröffentlichungen, die am Tag der Ausstellungspriorität oder danach erfolgten, außer Betracht bleiben.

Der erfinderische Schritt ist, ähnlich wie die erfinderische Tätigkeit im Patentrecht, jeweils im Einzelfall zu prüfen. Die frühere Ansicht, dass der Maßstab an die Erfindungshöhe, also der erfinderische Schritt beim

Gebrauchsmuster, im Allgemeinen geringer sei als die erfinderische Tätigkeit beim Patent, kann in Deutschland durch die Entscheidung „Demonstrationsschrank" des BGH (veröffentlicht u.a. in BGHZ 168, 142 und in Gewerblicher Rechtsschutz und Urheberrecht (GRUR) 2006, 842) als überholt angesehen werden. Es kann daher nicht generell gesagt werden, dass eine Erfindung, die nicht ganz „patentwürdig" ist, gebrauchsmusterfähig ist.

Im Gegensatz zum Patentrecht können in Deutschland (anders in Österreich) Verfahren nicht durch Gebrauchsmuster geschützt werden.

Das Patent- und Markenamt prüft bei einem Gebrauchsmuster nicht die sachlichen Voraussetzungen. Liegen die formellen Kriterien vor, so wird das Gebrauchsmuster in der Regel in das Gebrauchsmusterregister eingetragen (§ 8 GebrMG). Lediglich bei offensichtlich nicht dem Gebrauchsmusterschutz zugänglichen Gegenständen, z. B. Verfahren, erfolgt keine Eintragung. Dadurch wird ein schnelles Eintragungsverfahren erreicht, so dass der Inhaber aus dem Gebrauchsmuster sehr schnell Rechte geltend machen kann, ohne ein evtl. langwieriges Patenterteilungsverfahren abwarten zu müssen.

Eine professionell, d.h. von einem Patentanwalt eingereichte Gebrauchsmusteranmeldung ist in der Regel innerhalb von etwa 3 Monaten in das Gebrauchsmusterregister eingetragen. Im Vergleich dazu erstellt das DPMA bei einer Patentanmeldung nach etwa 8 Monaten einen ersten, sachlichen Bescheid, so dass bis zur Erteilung in der Regel mindestens 18 Monate vergehen.

Die sachlichen Kriterien werden erst bei einem Verletzungsverfahren durch das Zivilgericht oder im Löschungsverfahren vom Deutschen Patent- und Markenamt bzw. Bundespatentgericht geprüft.

Das Löschungsverfahren kann in Deutschland von jedermann durch einen Antrag auf Löschung eines Gebrauchsmusters beim Deutschen Patent- und Markenamt (DPMA) in Gang gesetzt werden. Somit besteht im Gegensatz zum Patent eine „doppelte" Verteidigungsmöglichkeit gegen ein Gebrauchsmuster (Zivilgerichte einerseits, DPMA und Bundespatentgericht andererseits). [134]

Kosten

Gebührenart	Euro
Anmeldegebühr	40,00 Euro
Recherchegebühr (für Eintragung nicht erforderlich)	250,00 Euro
1. Aufrechterhaltungsgebühr nach 3 Jahren	210,00 Euro
2. Aufrechterhaltungsgebühr nach 6 Jahren	350,00 Euro
3. Aufrechterhaltungsgebühr nach 8 Jahren	530,00 Euro
Löschungsantrag	300,00 Euro

Quelle: DPMA Stand September 2013

Da man beim Gebrauchsmusterschutz ebenfalls eine Patentnummer erhält, die man in Prospekten und im Schriftverkehr angeben kann, hat alleine dies für viele Unwissende eine hohe Abschreckungswirkung. Daher wäre dieser Weg für Start-ups eine günstige Möglichkeit die eigenen Erzeugnisse zumindest einigermaßen abzusichern.

Der Schutz greift bereits mit der Anmeldung. Produkte können also mit dem Zusatz „Gebrauchsmuster angemeldet" oder auf Englisch „Patent pending" versehen werden.

Ausgewählte Patente

Inhaltsverzeichnis

3D Drucker

Die Wiedergabe der Patentschriften erfolgt aus Platzgründen gekürzt. Anhand der **Anmeldenummer** lassen sich weiterführende Informationen recherchieren.

Inkjet/Sanddrucken/Pulverdrucken

26 Ausgewählte Patente

Zusammensetzung eines Bindergels

Veröffentlichungsnummer	US5660621 A
Publikationstyp	Erteilung
Anmeldenummer	US 08/581,319
Veröffentlichungsdatum	26. Aug. 1997
Eingetragen	29. Dez. 1995
Prioritätsdatum	29. Dez. 1995
Gebührenstatus	Verfallen
Auch veröffentlicht unter	US5851465, WO1997026302A1
Erfinder	James F. Bredt
Ursprünglich Bevollmächtigter	Massachusetts Institute Of Technology

Patentzitate (64), Nichtpatentzitate (10), Referenziert von (29),Klassifizierungen (20), Legal Events (7)

Externe Links: USPTO, USPTO-Zuordnung, Espacenet

Das Patent beschreibt die chemische Zusammensetzung für einen Binder für Pulverdrucker nach dem Inkjet Prinzip. Das Patent ist verfallen.

DETAILED DESCRIPTION OF THE INVENTION

To make the binder composition according to the preferred embodiment, the following components are combined and mixed thoroughly to dissolve the solids:

distilled water 385.9 cc (385.9 g)propylene glycol 58.4 cc (65.1 g)triethanolamine 21.7 cc (24.4 g)diethylene glycol monobutyl ether 12.6 cc (12.2 g)polyethylene glycol 1.0 gthymol blue 0.5g

To this mixture, 525.0 cc (735.0 g) of Nyacol 9950 are added. The silica appears to flocculate upon mixing. According, this mixture should be allowed to stand for a time or should be filtered by pumping in a closed circuit through a 5 µm filter for a time to redisperse the flocs.

The specific gravity of this mixture is 1.21 for 17.5 vol. % silica. The pH is between 9 and 9.5. The viscosity is approximately 2 to 3 cP (0.002 to 0.003 Pa-s). The surface tension is 54 dyn/cm (0.054 Pa-m).

Other methods to reduce the pH of the binder composition to cause gelation are possible. For example, gaseous CO_2 can be applied to each layer of the powder after printing of the binder.

The invention is not to be limited by what has been particularly shown and described except as indicated by the appended claims.

Selektiver Pulverdrucker mit Materialmanagenment

Veröffentlichungsnummer	US8523554 B2
Publikationstyp	Erteilung
Anmeldenummer	US 13/150,913
Veröffentlichungsdatum	3. Sept. 2013
Eingetragen	1. Juni 2011
Prioritätsdatum	2. Juni 2010
Erfinder	Ya Ching Tung, Kwo Yuan Shi
Ursprünglich Bevollmächtigter	Microjet Technology Co., Ltd.

Klassifizierungen (10), Legal Events (1)

Externe Links: USPTO, USPTO-Zuordnung, Espacenet

Kommentar:

Die Anmelder stellen einen kompletten Pulverdrucker vor, der die bekannten Probleme (Staub,Verbrauch, Rückgewinnungsverluste) beim Materialhandling zumindest reduzieren soll. Das Gerät erreicht Ausmaße die dennoch eher einem großen Kopierer als einem Desktopgerät entsprechen, sehen ihre Erfindung jedoch als Bürogerät an.

Mehrere Maßnahmen sollen den Staubaustritt minimieren, somit die Umgebung davor verschonen, den Anwender entlasten und auch die Fertigungszeit reduzieren.

Wesentliche Komponente ist die Kartusche(1) aus der das Pulver(a) gefördert wird. Sie verfügt über ein System aus einer Walze(121) mit Aussparungen, sowie rotierenden Förderschaufeln (112+113), die das Pulver(a) innerhalb der Kartusche zu verschiedenen Zonen befördern können. Wesentliches Element ist hierbei diese speziell geformte Walze. Sie dosiert die Menge des Pulvers, durch ihre Stellung. So lange sie sich dreht wird Pulver aufgenommen und durch eine nach unten gerichtete Öffnung ausgeschüttet. Gelangt das Positionierungssystem an eine Stelle, an der später kein Druck erfolgen soll, dreht sich die Walze in eine Stellung die die Auschüttungsöffnung verschließt. Auf diese Weise werden nur Bereiche mit Pulver versorgt, auf denen später auch gedruckt wird. Somit reduziert sich die spätere Entfernung des überschüssigen Materials. Eine an sich gute Idee. Die Dosierwalze(121) besitzt 3 unterschiedlich tief ausgefräste Kavitäten(X,Y,Z). Dies soll eine gleichmäßige Verteilung des Pulvers auf das Druckbett gewährleisten. Weitere Elemente des Pulverdruckers und dessen Materialhandlings werden detailliert beschrieben, hierzu bitte das vollständige Dokument sichten dort sind auch weitere Zeichnungen enthalten.

Zeichnungen:

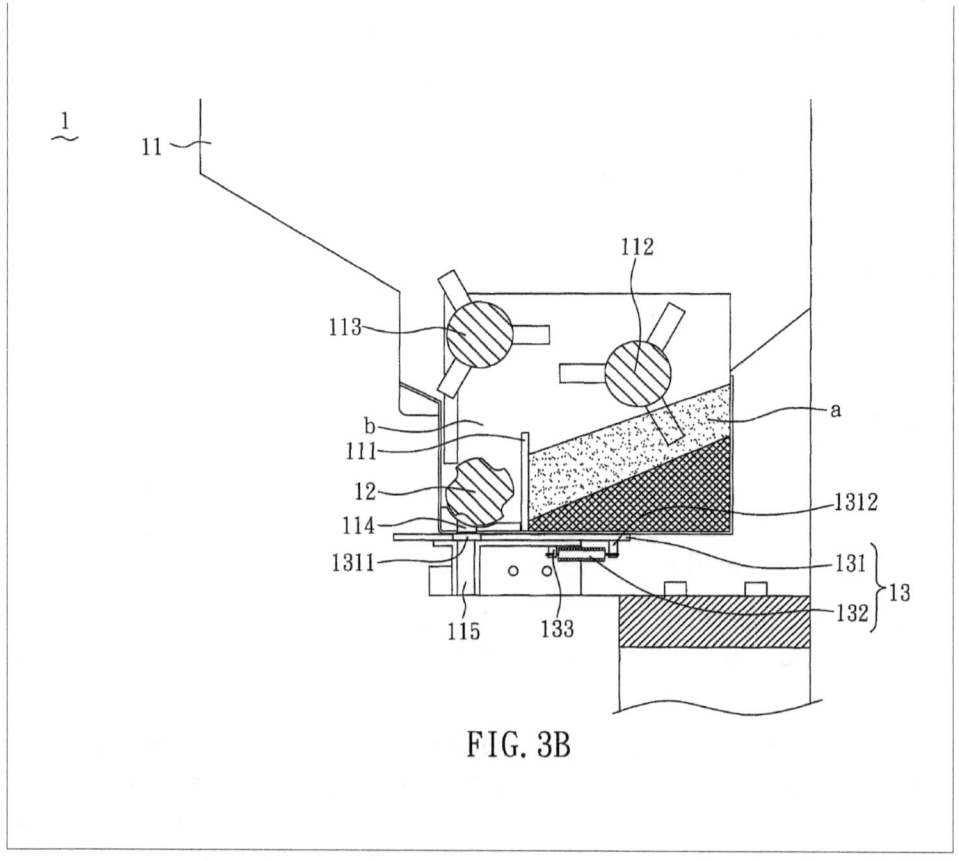

FIG. 3B

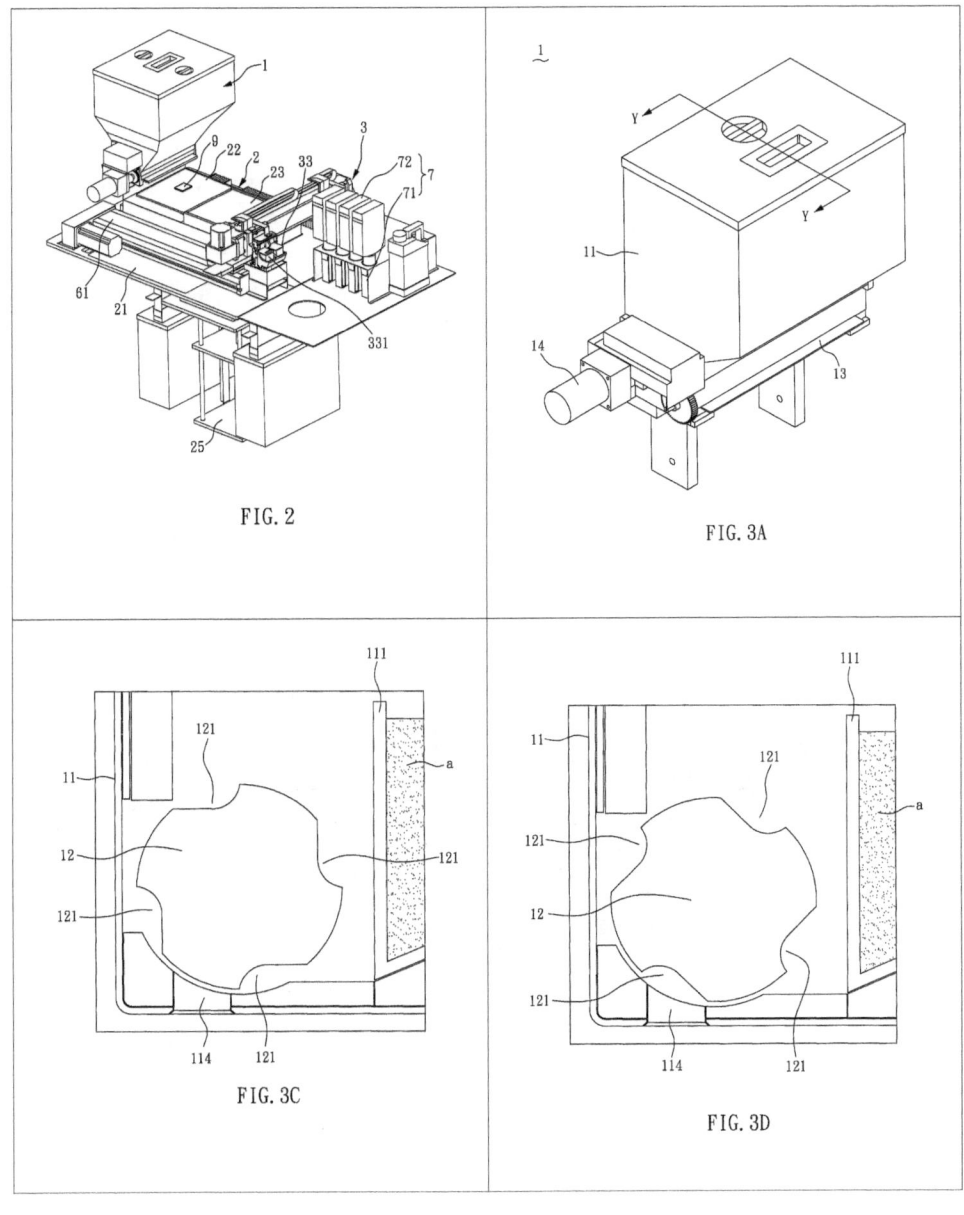

FIG. 2

FIG. 3A

FIG. 3C

FIG. 3D

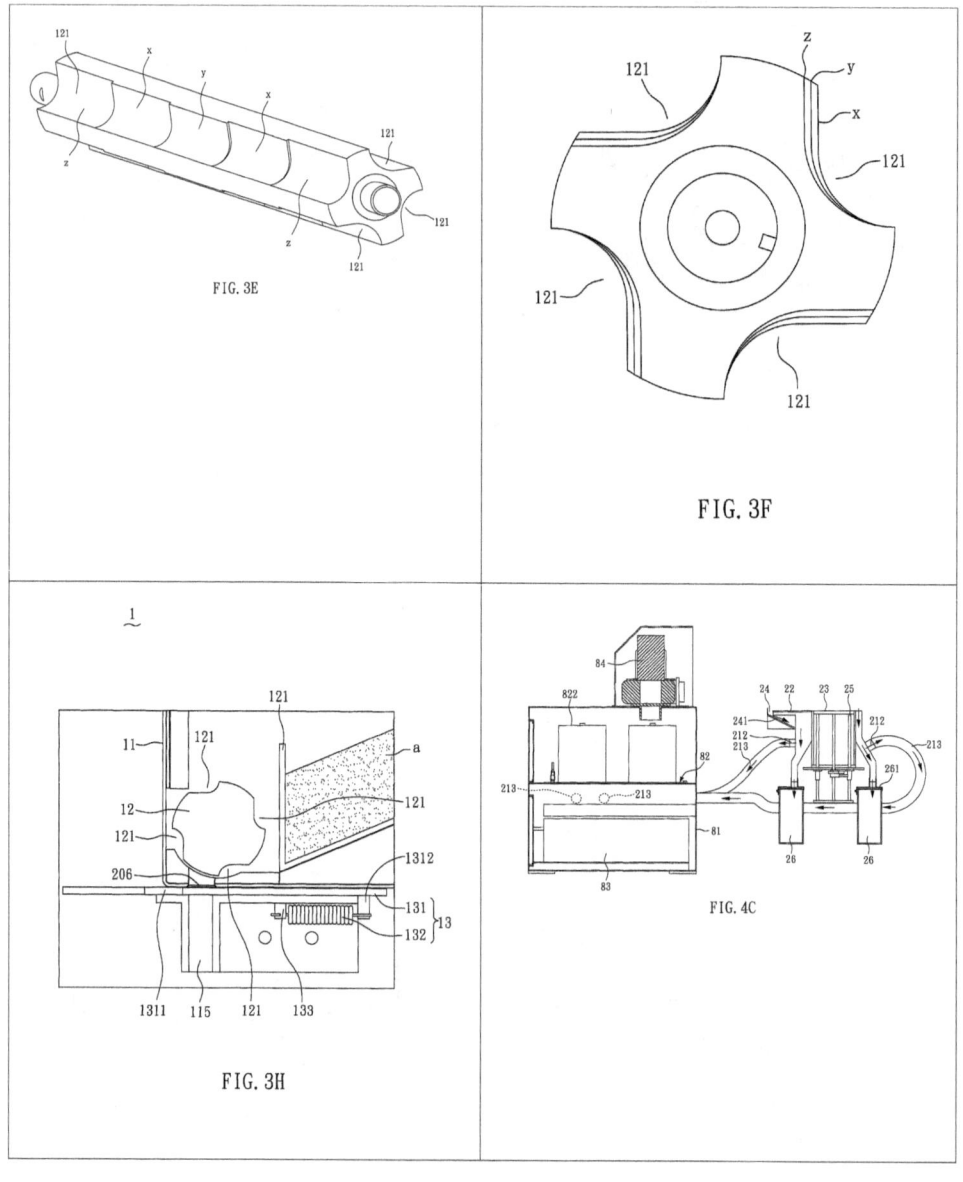

FIG. 3E

FIG. 3F

FIG. 3H

FIG. 4C

SUMMARY OF THE INVENTION

The object of the present invention is to provide a three-dimensional object-forming apparatus, which has a quantitative powder-supplying tank system to regulate the corresponding times between the cavities of the in-batches rationing roller under rolling and the dropping-powder opening according to requirements of different powder-application thicknesses so as to control the output amount of the construction powder. Therefore, redundant construction powder drawn in the powder collection tank can be reduced to avoid the waste of the construction powder and decrease the production costs. In addition, each cavity of the in-batches rationing roller has a plurality of compartments, and the capacity of the compartments increases from the center of the cavities to the both sides thereof so as to achieve even powder application and improve the drawback of powder deficiency at the both sides.

Besides, the three-dimensional object-forming apparatus of the present invention further comprises a heating device used to heat during the printing of the printing module to accelerate the combination between the adhesive and the construction powder and reduce one-third to half time of forming a three-dimensional object. The three-dimensional object-forming apparatus of the present invention further comprises a successive liquid-supplying device which can successively supply an adhesive into the printing cartridge to make the printing module inkjet-print on the construction powder for a long term of time.

Furthermore, the three-dimensional object-forming apparatus of the present invention has a dust-proof device for a driving component to prevent the contamination of the disturbed powder during the powder application and inkjet printing so that the apparatus and components of the three-dimensional object-forming apparatus all can be kept anytime in a normal operation and achieve absolute dust-proofing overall.

Meanwhile, the three-dimensional object-forming apparatus of the present invention is provided with an inkjet-print head maintenance device which comprises a cleaning unit and a sealing unit. After the inkjet-printing operation is completed by the inkjet-print head, the inkjet-print head can be completely cleaned by the scrapers of the cleaning unit and sealed in the sealing part of the sealing unit to achieve thorough anti-contamination and anti-drying of the inkjet-print head.

Moreover, the three-dimensional object-forming apparatus of the present invention has the design of the print quality detection, in which ground glass is used as a print quality detection member to real-timely observe whether the pattern inkjet-printed by the inkjet-print head is normal and determine whether the inkjet-print head is blocked so as to clean the inkjet-print head in time and keep the print quality.

In order to achieve the abovementioned objects, a generalized aspect of the present invention provides a three-dimensional object-forming apparatus comprising an in-batches powder-rationing tank system, a construction tank system, a printing powder-applying system, a rapid drying heating system, a printing maintenance device, a dust-proof device, a successive liquid-supplying device, a powder auto-filtrating and recycling device, and a print quality detection device.

BRIEF DESCRIPTION OF THE DRAWINGS

FIG. 1 shows an exterior view of the three-dimensional object-forming apparatus in the preferred example of the present invention;

FIG. 2 shows an interior view of the three-dimensional object-forming apparatus in the preferred example of the present invention;

FIG. 3A shows a structural view of the quantitative powder-supplying tank system;

FIG. 3B shows a Y-Y cross-sectional view of FIG. 3A;

FIG. 3C shows a structural view of the partial powder-supplying tank and in-batches rationing roller in FIG. 3B;

FIG. 3D shows a structural view of supplying powder in FIG. 3B;

FIG. 3E shows a structural view of the in-batches rationing roller in FIG. 3B;

FIG. 3F shows a front view of the in-batches rationing roller in FIG. 3B;

FIG. 3G is a structural view of the closing device and dropping-powder channel shown in FIG. 3B;

FIG. 3H shows a structural view of the opening of the closing device unconnected to the dropping-powder opening shown in FIG. 3G;

FIG. 4A shows a structural view of the construction tank system;

FIG. 4B shows a structural view of the partial remaining powder auto-collection area of the construction tank system;

FIG. 4C shows a view of recycling the remaining powder in the construction tank system;

FIG. 5A shows a structural view of the printing powder-applying system;

FIG. 5B shows a structural view of the printing module of the printing powder-applying system;

FIG. 5C shows a cross-sectional view of FIG. 5B;

FIG. 6 shows a view of the dust-proof device;

FIG. 7A shows a view of the printing maintenance device;

FIG. 7B shows a structural view of the cleaning unit;

FIG. 7C shows a cross-sectional view of FIG. 7B;

FIG. 7D shows a structural view of the sealing unit;

FIG. 8 shows a view of the liquid supplying in the successive liquid-supplying device;

FIG. 9 shows a view of the connection between the powder auto-filtrating and recycling device and the three-dimensional object-forming apparatus; and

FIG. 10 shows a cross-sectional view of the powder auto-filtrating and recycling device.

DETAILED DESCRIPTION OF THE PREFERRED EMBODIMENT

Several typical embodiments showing the features and advantages of the present invention are explained in relation in the following paragraphs, and it is to be understood that many other possible modifications and variations can be made without departing from the spirit and scope of the invention as hereinafter claimed.

With reference to FIGS. 1 and 2, they are

exterior and interior views of the three-dimensional object-forming apparatus in a preferred example of the present invention. As shown in FIGS. 1 and 2, the three-dimensional object-forming apparatus of the present invention mainly includes an in-batches powder-rationing tank system 1, a construction tank system 2, a printing powder-applying system 3, a rapid drying heating system 4 (shown in FIG. 5C), a printing maintenance device 5 (shown in FIG. 7A), a dust-proof device 6 (shown in FIG. 6), a successive liquid-supplying device 7 (shown in FIG. 8), a powder auto-filtrating and recycling device 8 (shown in FIG. 9), and a print quality detection device 9.

The in-batches powder-rationing tank system 1 and the construction tank system 2 of the present invention are provided in view of that there is no in-batches powder-rationing device in the conventional rapid-forming apparatus and it causes the uneven density and redundant powder drawn in the trihedral auto-recycling tub resulting in uneven powder application. Therefore, an in-batches rationing roller and a trihedral auto-recycling tub are installed in the powder-supplying system to overcome the abovementioned drawbacks. How to overcome the drawbacks is the main topic of developing the in batches powder-rationing tank system 1 and the construction tank system 2 of the present invention. The following are illustrations of the related components.

With reference FIGS. 3A and 3B, FIG. 3A shows a structural view of the in-batches powder-rationing tank system in a preferred example of the present invention, and FIG. 3B shows a Y-Y cross-sectional view of FIG. 3A. As shown in FIGS. 3A and 3B, the in-batches powder-rationing tank system 1 includes at least one powder-supplying tank 11, an in-batches rationing roller 12, and a closing device 13. The powder-supplying tank 11 is a hollow tank structure and used for storage of the construction powder "a". Within the powder-supplying tank 11, a baffle plate 111, a first roller 112, and a second roller 113 are installed. Additionally, a dropping-powder opening 114 and a dropping-powder channel 115 are disposed on the bottom of the powder-supplying tank 11. A lateral of the baffle plate 111 and the dropping-powder opening 114 are separated by a dropping-powder zone "b". The construction powder "a" accumulated outside the dropping-powder zone "b" of the baffle plate 111 can be disturbed by the first roller 112 and then drop within the dropping-powder zone "b" of the baffle plate 111 by the rotation of the second roller 113.

With reference to FIGS. 3C and 3D, they are partially structural views of the powder-supplying tank and the in-batches rationing roller in FIG. 3B. As shown in FIGS. 3C and 3D, the in-batches rationing roller 12 is installed in the dropping-powder zone "b" of the powder-supplying tank 11, close to the dropping-powder opening 114, used to supply the construction powder "a" in batches required for total application of a construction-forming area, and has a plurality of cavities 121. Each cavity 121 is mainly used to receive the construction powder "a". When the cavities 121 of the in-batches rationing roller 12 do not correspond to the dropping-powder opening 114, the construction powder "a" can not be output (as shown in FIG. 3C). On the contrary, when one of the cavities 121 corresponds to the dropping-powder opening 114, the construction powder "a" contained in the powder-supplying tank 11 are output via the dropping-powder opening 114 (as shown in FIG. 3D).

Besides, in the in-batches powder-rationing tank system 1 of the present invention, the corresponding times between the cavities 121 of the in-batches rationing roller 12 under rolling and the dropping-powder opening 114 can be regulated by a motor 14 according to the requirements of different powder application thicknesses so as to control the output amount of the construction powder "a" to avoid the waste of the construction powder "a". For example, if the powder application thickness of the construction-forming area has the maximum of 0.12 mm and the minimum of 0.08 mm. The amount of the construction powder "a" received in a cavity 121 of the in-batches rationing roller 12 approximately forms a thickness of 0.04 mm. Therefore, when the construction powder "a" is formed in a thickness of 0.08 mm, the motor 14 has to rotate twice to make two cavities 121 of the in-batches rationing roller 12 connect to the dropping-powder opening 114 and thus the construction powder "a" received in the cavities 121 can be output via the dropping-powder opening 114. When the construction powder "a" is formed in a thickness of 0.12 mm, the motor 14 has to rotate three times to make three cavities 121 of the in rationing roller 12 cannot to the dropping-powder opening 114 and thus the construction powder "a" received in the cavities 121 can be output via the dropping-powder opening 114. Accordingly, the redundant construction powder "a" drawn into a powder collection tank can be reduced.

With reference to FIGS. 3E and 3F, they are structural and front views of the in-batches rationing roller shown in FIG. 3B. As shown in FIGS. 3E and 3F, each cavity 121 of the in-batches rationing roller 12 of the present invention has a plurality of compartments "x", "y", and "z". In the present example, one compartment "x", two compartments "y", and two compartments "z" are contained in each cavity 121, but not limited thereto. The compartment "x" is set in the center of the cavities 121 and both sides of the compartment "x" are provided respectively with the compartments "y". The compartments "z" are set at the other side of the compartments "y". The cavities of the compartment "x" are shallowest and have the least amount of the received powder. The cavities of the compartments "y" are deeper and have more amount of the received powder than those of the compartment "x". The cavities of the compartments "z" are deeper than those of the compartment "x" and the compartments "y" and thus have the largest amount of the received powder. In other words, the amount of the received powder in one compartment "x" and plural compartments "y" and "z" increase from the center to the both sides of the cavities 121, i.e. compartment "x"<compartments "y"<compartments "z". Based on the structural designs that each cavity 121 has one compartment "x" and plural compartments "y" and "z" and the capacity of one compartment "x" and plural compartments "y" and "z" increases from the center of the cavities 121 to the both sides thereof, the construction powder "a" can be applied evenly on the construction-forming area so as to overwhelm the drawbacks of more and more differences of the construction powder amounts between the center and the both sides in the conventional technique as the times of the powder application increase.

With reference to FIG. 3G, it is a structural view of the closing device and dropping-powder channel shown in FIG. 3B. As shown in FIG. 3G, the closing device 13 included in the in-batches powder-rationing tank system 1 of the present invention has a board 131, an elastic member 132, and a retention member 133. The board 131 is movable and has an opening 1311 and a fixing member 1312. An end of the elastic member 132 is connected to the fixing member 1312, and the other end thereof is connected to the retention member 133 mounted on the bottom of the powder-supplying tank 11. During the powder supply of the powder-supplying tank 11, the board 131 of the closing device 13 is moved by a thrust towards the direction "f" and thus the opening 1311 thereof is connected to the dropping-powder opening 114. At this instance, the construction powder "a" received in one cavity 121 of the in-batches rationing roller 12 is output via the dropping-powder opening 114, the opening 1311, and the dropping-powder channel 115 (as shown in FIG. 3G).

Stereolithographie

Verfahren um Stützkonstruktionen überflüssig zu machen

(19) **JAPANESE PATENT OFFICE**

PATENT ABSTRACTS OF JAPAN

(11) Publication number: **2002046188 A**

(43) Date of publication of application: **12.02.02**

(51) Int. Cl **B29C 67/00**

(21) Application number: **2000267741**

(22) Date of filing: **01.08.00**

(71) Applicant: **UNIRAPID INC**

(72) Inventor: **KENMOKU MASATAKA**

(54) **OPTICAL SHAPING METHOD FOR REUTILIZING RESIN AND DISPENSING WITH SUPPORT**

(57) Abstract:

PROBLEM TO BE SOLVED: To reutilize a cured photosetting resin and to shape a material without adding a support even in a shape requiring the support.

SOLUTION: A material prepared by mixing a cured resin powder with a liquid resin is used to enable shaping without adding the support even in a shape requiring the support.

COPYRIGHT: (C)2002,JPO

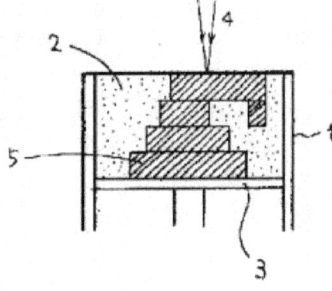

Kommentar vom Autor: Japanisches Patent mit cleverem Ansatz, falls dieses Verfahren funktioniert. Durch Verwendung des in vorherigen Druckdurchgängen angefallenen und bereits ausgehärtetem Material, können die sonst üblich Stützkonstruktionen entfallen. Dies reduziert die Vorbereitungs- und Nachbearbeitungszeit und vereinfacht die Anwendung von SL Druckern stark für Laien.

Ablauf: Bereits ausgehärtetes Material wird pulverisiert und mit dem flüssigen Harz gemischt, so dass eine Paste entsteht.

Wenn, wie in der Abbildung zu sehen auskragende Überhänge im Objekt gedruckt werden müssen, so können diese nicht mehr herunterfallen, da sie von den Feststoffpartikeln in der Paste daran gehindert werden.

Allerdings erfordert dieses einfache und umweltschonende Trick einen modifizierten Druckablauf nach dem Verständnis des Verfassers, denn die Paste müsste ja auch nach dem Absenken jeder Schicht neu, auf die oberste Lage neu aufgetragen werden.

Die Erklärungen hierzu sind auf japanisch verfasst, so dass man – sofern man dies nicht übersetzen kann - auf die verbleibenden Abbildungen zum Verständnis zurück greifen muss.

Zeichnungen:

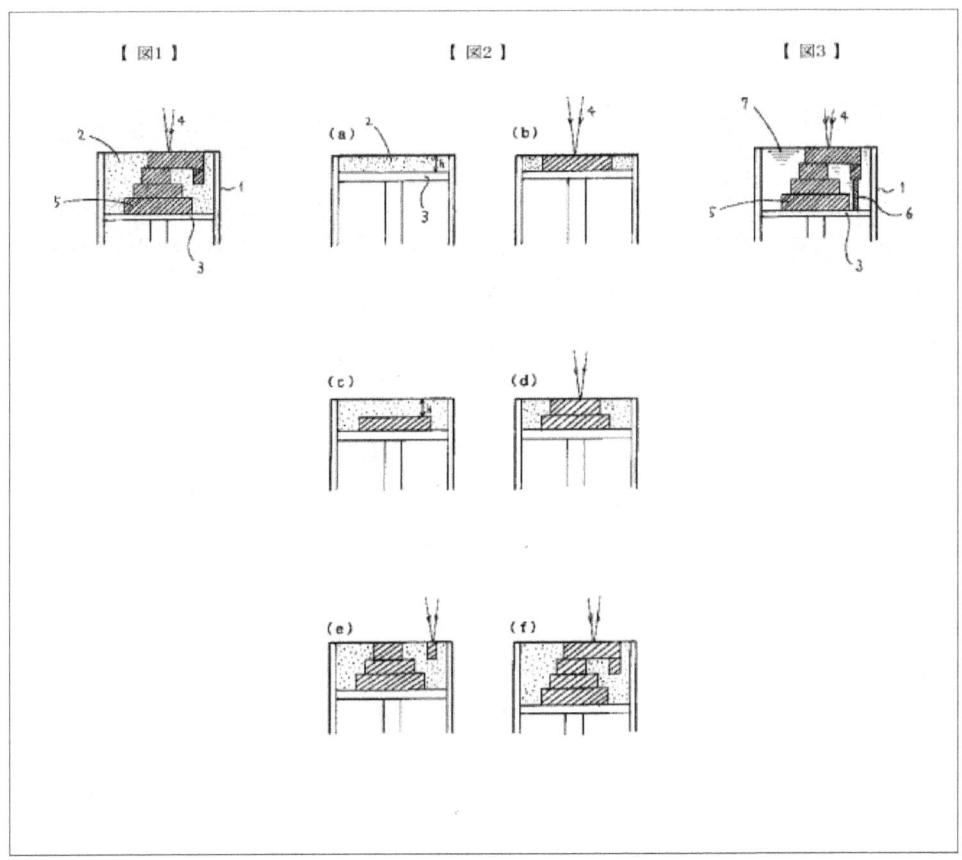

Hochtransparente lichthärtende Harzzusammensetzung

Veröffentlichungsnummer	US8377623 B2
Publikationstyp	Erteilung
Anmeldenummer	US 12/745,036
Veröffentlichungsdatum	19. Febr. 2013
Eingetragen	21. Nov. 2008
Prioritätsdatum	27. Nov. 2007
Auch veröffentlicht unter	EP2215525A1, EP2215525A4,US20100 304100, WO2009070500A1
Erfinder	John Wai Fong
Ursprünglich Bevollmächtigter	3D Systems, Inc.

Patentzitate (25), Nichtpatentzitate (1), Klassifizierungen (13),Legal

40 Ausgewählte Patente

> **Kommentar vom Autor:**
>
> Die Erfindung beschreibt die chemische Zusammensetzung eines hochtransparenten, lichtaushärtenden Harzes, das zur Herstellung von glasklaren Modellen verwendet werden kann.
>
> Hierbei wird darauf verwiesen, dass man unter transparent nicht bisher erreichte Lichtdurchlässigkeiten versteht, sondern eben eine Transparenz die die von Glas entspricht. Es sind US und Europatente vorhanden. Das Europatent nur in englischer und französischer Sprache. Nachfolgend einige Auszüge, die sehr detailliert, aber auch nur für Fachleute der entsprechenden Bereiche in der Chemie verständlich sein dürften.

ANSPRÜCHE (OCR-Text kann Fehler enthalten)

What is claimed is:

1. A photocurable composition comprising: (a) 35-80% by weight of a cationically curable component comprising a polyglycidyl epoxy compound;

(b) 5-60% by weight of a free radically active component comprising an ethoxylated and/or propoxylated poly(melh)acrylate;

(c) 0.1-10% by weight of a antimony-free cationic photoinitiator; (d) 0.01-10% by weight of a free radical photoinitiator; and

(e) 0-40% by weight of one or more optional components wherein the percent by weight is based on the total weight of the photocurable composition and wherein the photocurable composition, after cure, is clear and has a yellowness index/inch thickness of less than 70 and a flexπral modulus of at least 1000 MPa. 2. The photocurable composition of claim 1 wherein the cationically curable component and free radically active component produce a C:H:O ratio of at least 3.0:3.75:1

3. The photocurable composition of claim 2 wherein the C:H:O ratio is at least 3.5:5.0:1.

4. The photocurable composition of claim 2 wherein the poly(meth)acrylate further comprises a non-aromatic poly(meth)acrylate. 5. The photocurable composition of claim 4 wherein the poly(meth)acry late is an alicyclic poly(meth)acrylale.

6. The photocurable composition of claim 1 wherein the antimony-free cationic photoinitiator is of the formula (1):

where

R1, R2 and R3 are each independently of one another Cβ-is aryl that is unsubstituted or substituted by radicals selected from the group consisting of C1-6 alkyl; C\.6 alkoxy; Ci .6 alkylthio; halogen; amino groups; cyano groups; nitro groups, and arylthio;

Q is boron or phosphorus;

X is a halogen atom; and m is an integer corresponding to the valence of Q plus 1.

7. The photocurable composition of claim 1 wherein the polyglycidyl epoxy compound comprises a hydrogenated bisphenol epoxy-containing compound.

8. The photocurable composition of claim 7 wherein the cationically curable component further comprises an oxetane compound.

9. The photocurable composition of claim 1 wherein the cationic photoinitiator is a triarylsulfonium hexafluorophosphate salt.

10. A photocurable composition comprising: (a) 40-70% by weight of a cationically curable component comprising one or more epoxy-containing compounds and one or more oxetane compounds;

(b) 15-60% by weight of a free radically active component comprising

(i) at least one ethoxylated or propoxylated poly(meth)acrylate or mixture thereof and

(ii) a non-aromatic poly(meth)acrylate;

(c) 0, 1-10% by weight of an antimony-free cationic photoiniliator;

(d) 0.1-10% by weight of a free radical photoiiiitiator;

(e) 0-40% by weight of one or more optional components wherein the amount of the ethoxylated or propoxylated poly(meth)acrylate or mixture thereof is greater than 40% by weight of the total amount of free radically active component and wherein the photocurable composition, after cure by exposure to actinic radiation and optionally heat, has a yellowness index/inch thickness of less than 70.

11. The photocurable composition of claim 10 wherein the cationically curable component and free radically active component produce a C:II:O ratio of at least 3.0:3.75:1.

12. The photocurable composition of claim 1 1 wherein the C:H:O ratio is at least 3.5:5.0:1.

13. The photocurable composition of claim 10 wherein the cationically curable component comprises a hydrogenated bisphenol epoxy-containing compound.

14. A process for producing a colorless three dimensional article comprising: (a) forming a first layer of the photocurable composition of claim 1 on a surface; (b) exposing the layer imagewise to actinic radiation to form an imaged cross- section, wherein the radiation is of sufficient intensity to cause substantial curing of the layer in the exposed areas;

(c) forming a second layer of the composition of claim 1 on the previously exposed imaged cross-section;(d) exposing the second layer from step (c) imagewise to actinic radiation to form an additional imaged cross-section, wherein the radiation is of sufficient intensity to cause substantial curing of the second layer in the exposed areas and to cause adhesion to the previously exposed imaged cross-section; and (e) repeating steps (c) and (d) a sufficient number of times in order to build up the three-dimensional article.

15. A three-dimensional medical article produced according to the process of claim 14.

16. The three-dimensional article of claim 15 wherein the article is a container, headlight, shade or decorative object. 17. A process for producing a three dimensional article by jet printing comprising the steps of:

(a) applying successive droplets of the photocorable composition of claim 1 at targeted locations on a substrate in accordance with a desired shape stored on a computer file; (b) exposing the droplets to electromagnetic radiation to cure the droplets in the exposed areas; (c) repeating steps (a) and (b) a sufficient number of times in order to build up the three dimensional article.

18. The process of claim 17 wherein the substrate comprises paper, textiles, tiles, printing plates, wallpaper, plastic, powder or paste. 19. The process of claim 18 wherein the photociirable composition is exposed to electromagnetic radiation pixel by pixel, line by line, layer by layer, after several layers have been formed, and/or after all layers have been formed.

20. The process of claim 19 wherein the electromagnetic radiation employed is UV light, microwave radiation, visible light, or laser beams. 21. A photociirable composition comprising:

(a) 30-55% by weight of a cationically curable component comprising a hydrogenated aromatic polyglycidyl epoxy compound;

(b) 5-60% by weight of a free radically active component comprising an ethoxylated and/or propoxylated poly(meth)acrylate; (c) 0.1-10% by weight of a antimony-free cationic photoinitiator;

(d) 0.01-10% by weight of a free radical photoinitiator; and

(e) 0-40% by weight of one or more optional components

(f) 5-25% of an oxetane compound having one oxetane ring wherein the percent by weight is based on the total weight of the photocurable composition and wherein the photocurable composition has a C:II:O ratio of at least 3.0:3.75:1 and after cure, is clear and has a yellowness index/inch thickness of less than 70 and a fiexural modulus of at least 1000 MPa.

22. The photocuiable composition of claim 21 wherein the C;II:O ratio is at least 3.5:5.0:1.

Fotohärtender Harz mit ABS Eigenschaften

Publication number	CA2620714 A1
Publication type	Application
Application number	CA 2620714
PCT number	PCT/EP2006/066264
Publication date	22 Mar 2007
Filing date	12 Sep 2006
Priority date	13 Sep 2005
Also published as	CN101263428A, EP1924887A1, US8227048,US201102 93891, WO2007031505A1
Inventors	Carole Chapelat, 6 More »
Applicant	3D Systems, Inc., 8 More »
	Classifications (6), Legal Events (2)

External Links: CIPO, Espacenet

Kommentar:

Ein recht kurzer Patenttext, der die Zusammensetzung eines fotohärtenden Harzes mit ABS Eigenschaften verspricht.

ABSTRACT

The present invention provides a clear, low viscosity photocurable composition including (i) a cationically curable compound (ii) an acrylate-containing compound (iii) a polyol-conyaining mixture (iv) a cationic photoinitiator and (v) a free radical photoinitiator. The photocurable composition can be cured using rapid prototyping techniques to form opaque-white three-dimensional articles having ABS-like properties.

CLAIMS

1) A photocurable composition comprising:

a. 30-80% by weight of an epoxy-containing compound;

b. 5-40% by weight of a polyfunctional (meth)acrylate;

c. 5-40% by weight of a polyol-containing

mixture comprising (1) at least one component of low to medium molecular weight which component contains at least one epoxy or alcohol functionality and (2) at least one polyol, which is different from compound (1) and has a higher molecular weight than compound (1) d. a cationic photoinitiator;

e. a free radical photoinitiator; and optionally f. one or more stabilizers wherein the percent by weight is based on the total weight of the photocurable composition.

2) The photocurable composition of claim 1 wherein component (1) is a polyol chosen amongst the following types: poly(oxytetramethylene) polyol, poly(oxypropylene) polyol, poly(oxyethylene) polyol, hydroxy-terminated polybutadiene or hydroxy-terminated polysiloxane.

3) The photocurable composition of claim 2 wherein the molar ratio of the poly(oxytetramethylene)polyol over the at least one other polyol is equal to or less than 25.

4) The photocurable composition of any preceding claim wherein the polyol of component (2) is a polyether polyol.

5) The photocurable composition of any preceding claim wherein the photocurable composition is a clear liquid.

6) The photocurable composition of any preceding claim which, after cure by exposure to actinic radiation is opaque.

7) The photocurable composition of any preceding claim which, after cure by exposure to actinic radiation is opaque -white that simulates ABS.

FDM Drucker

Mehrfachextruder mit nur einem Antriebsmotor

Veröffentlichungsnummer	US7604470 B2
Publikationstyp	Erteilung
Anmeldenummer	US 11/396,845
Veröffentlichungsdatum	20. Okt. 2009
Eingetragen	3. Apr. 2006
Prioritätsdatum	3. Apr. 2006
Gebührenstatus	Bezahlt
Auch veröffentlicht unter	CN101460050A, 5 weitere »
Erfinder	Benjamin N. Dunn, 3 weitere »
Ursprünglich Bevollmächtigter	Stratasys, Inc.

Kommentar:

Gezeigt wird das erteilte US-Patent, einer Vorrichtung, mit der sich mindestens zwei Filamente mit zwei Extruderdüsen über nur einen Antriebsmotor zuführen lassen.)

BRIEF DESCRIPTION OF THE DRAWINGS

FIG. 1 is a side view of an extrusion-based layered manufacturing system with a portion broken away to show an extrusion head of the present invention.

FIG. 2A is a front perspective view of the extrusion head having a toggle-plate assembly positioned in a build state.

FIG. 2B is a front perspective view of the extrusion head, where the toggle-plate assembly is positioned in a support state.

FIG. 3 is a left side view of the extrusion head 20, where the toggle-plate assembly is positioned in the build state.

FIG. 4A is an expanded view of a left-portion of the toggle-plate assembly shown in FIG. 2A.

FIG. 4B is an expanded view of a right-portion of the toggle-plate assembly shown in FIG. 2B.

FIG. 5 is a sectional view of section 5-5 taken in FIG. 3, showing the toggle-plate assembly.

FIG. 6A is a front exploded view of the extrusion head.

FIG. 6B is a rear exploded view of the extrusion head.

FIGS. 7A-7E are expanded views of a toggle bar of the extrusion head in use for positioning the toggle-plate assembly between the build state and the support state

Zeichnungen (8 von 15)

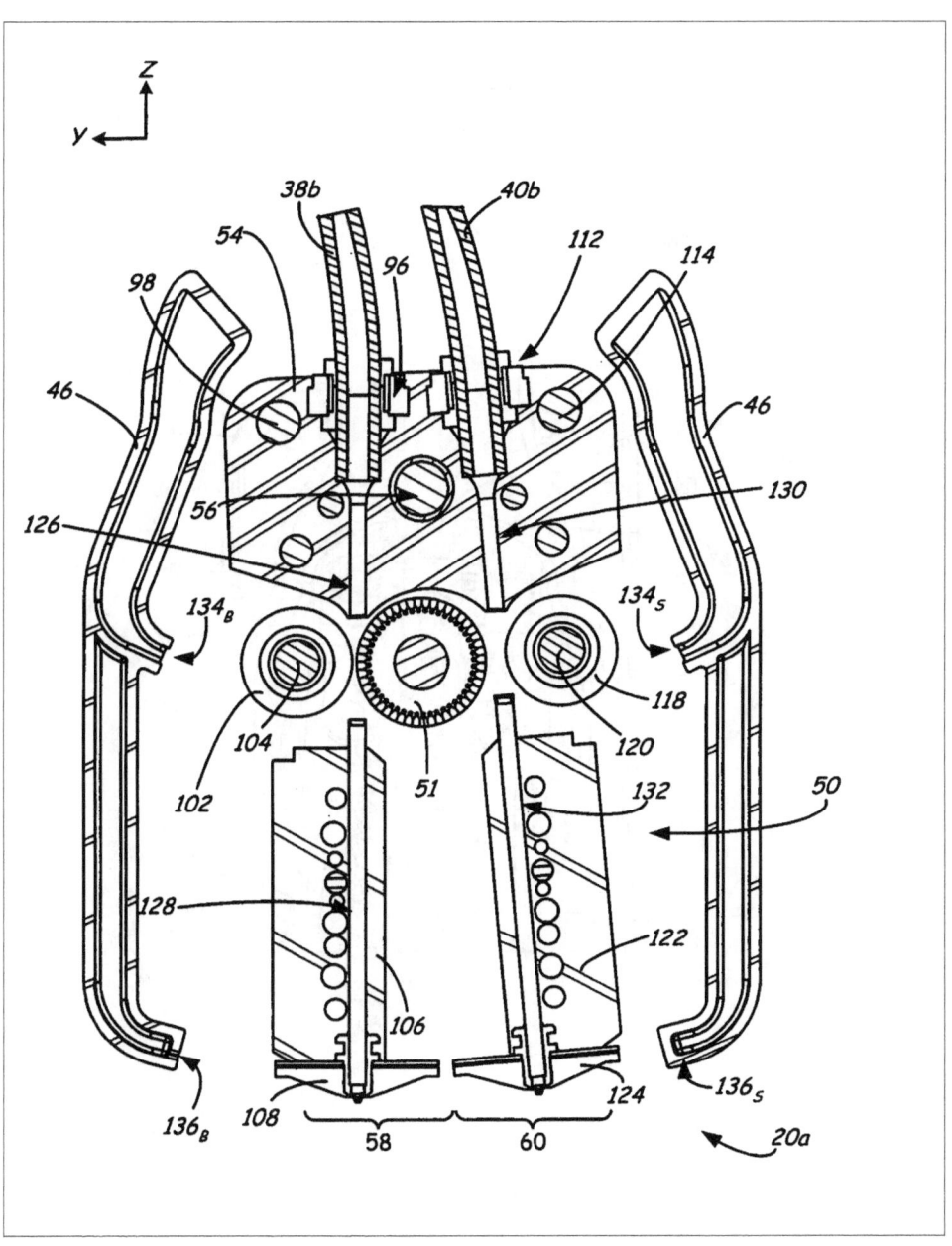

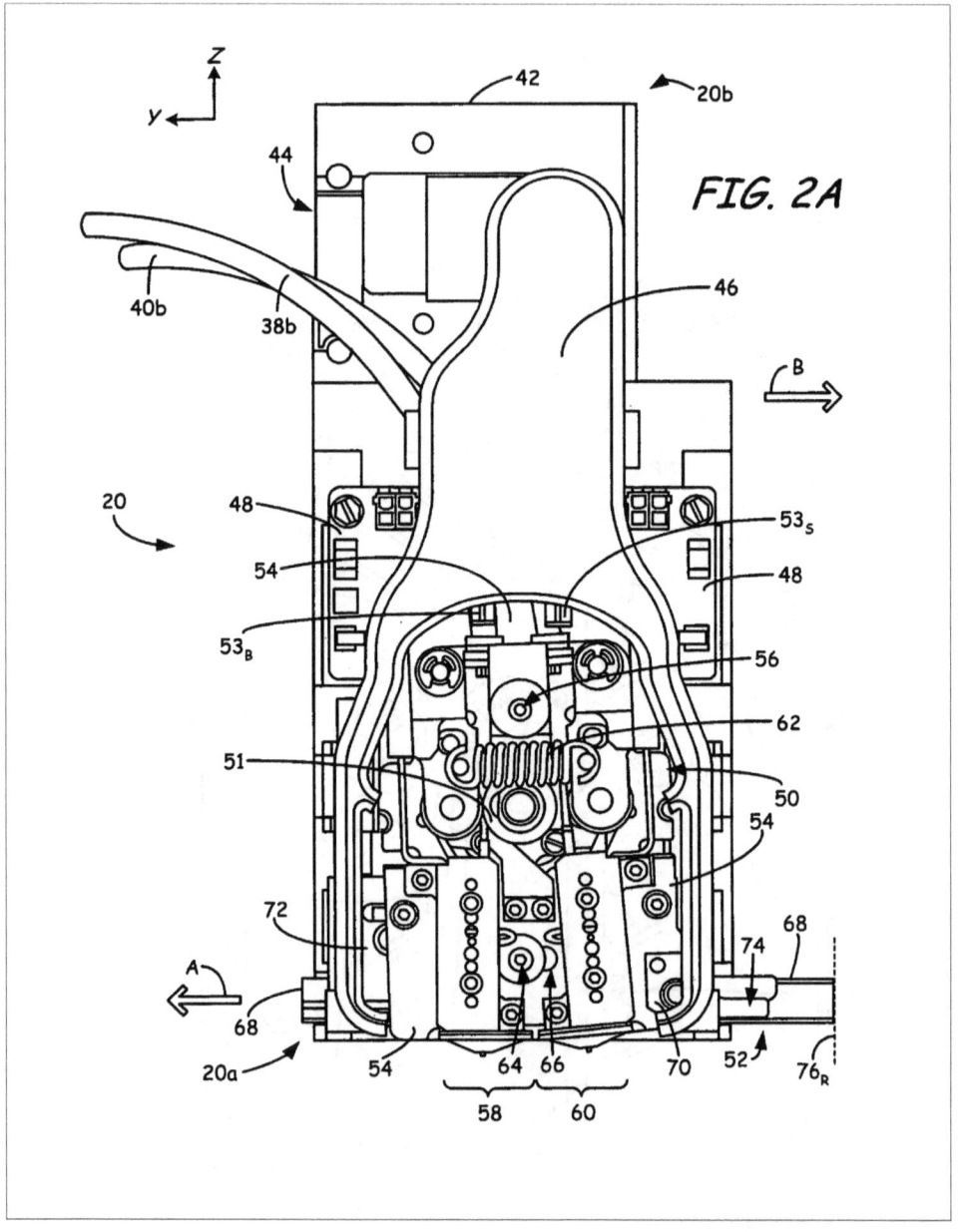

FIG. 2A

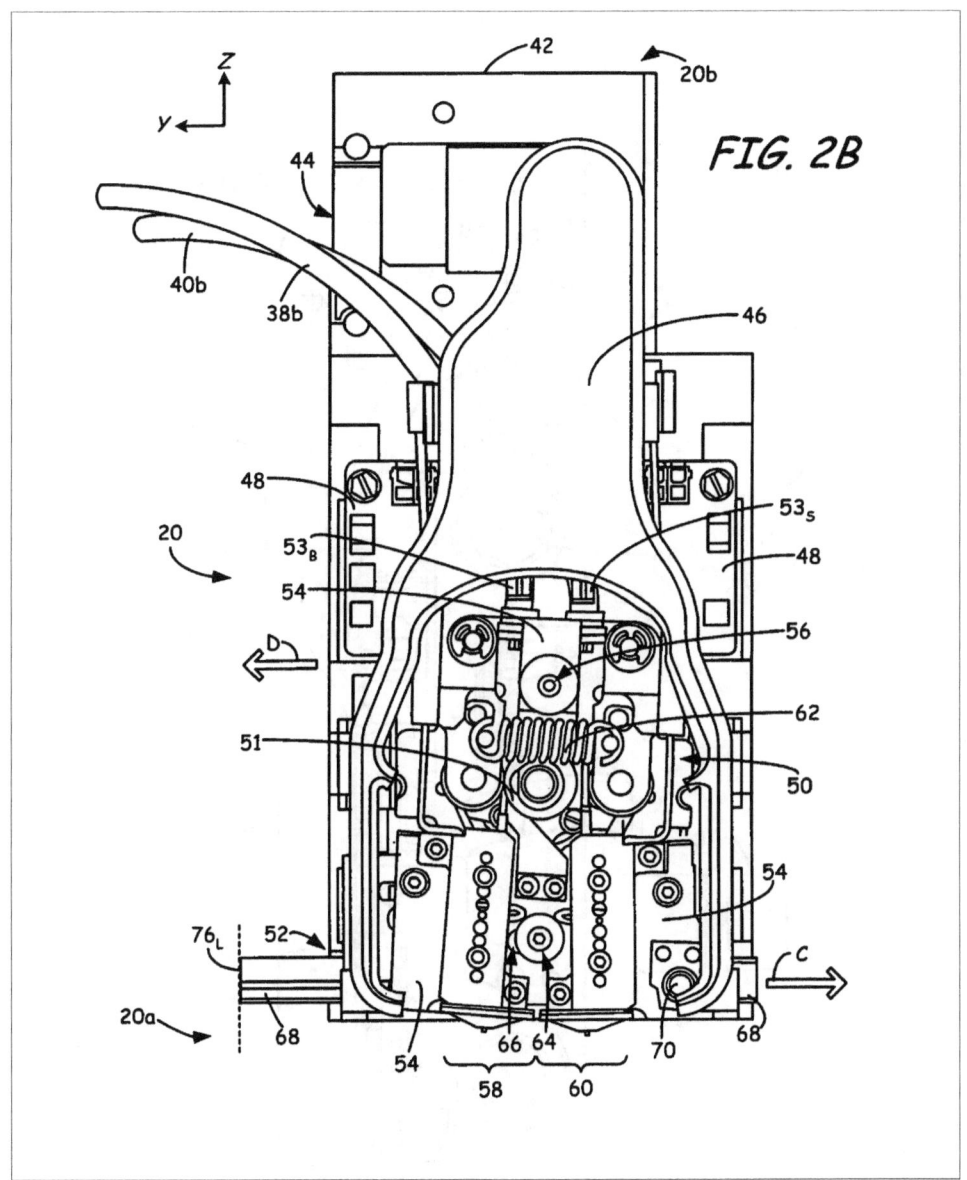

FIG. 2B

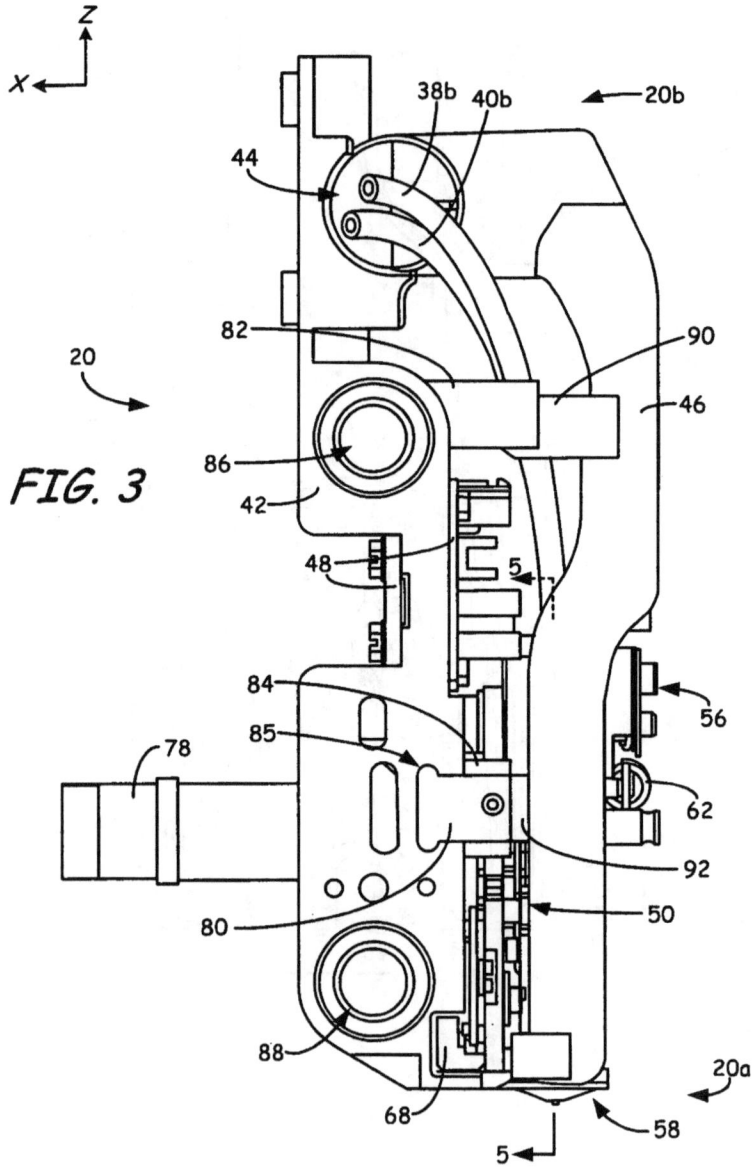

FIG. 3

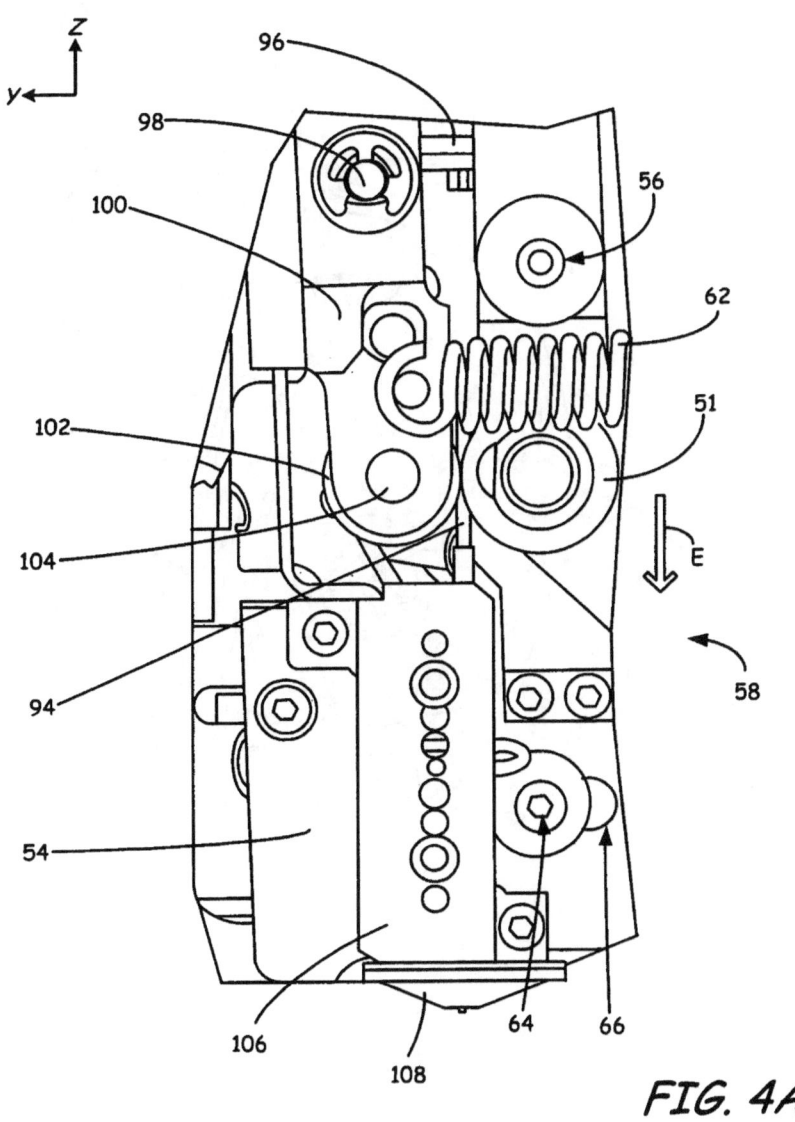

FIG. 4A

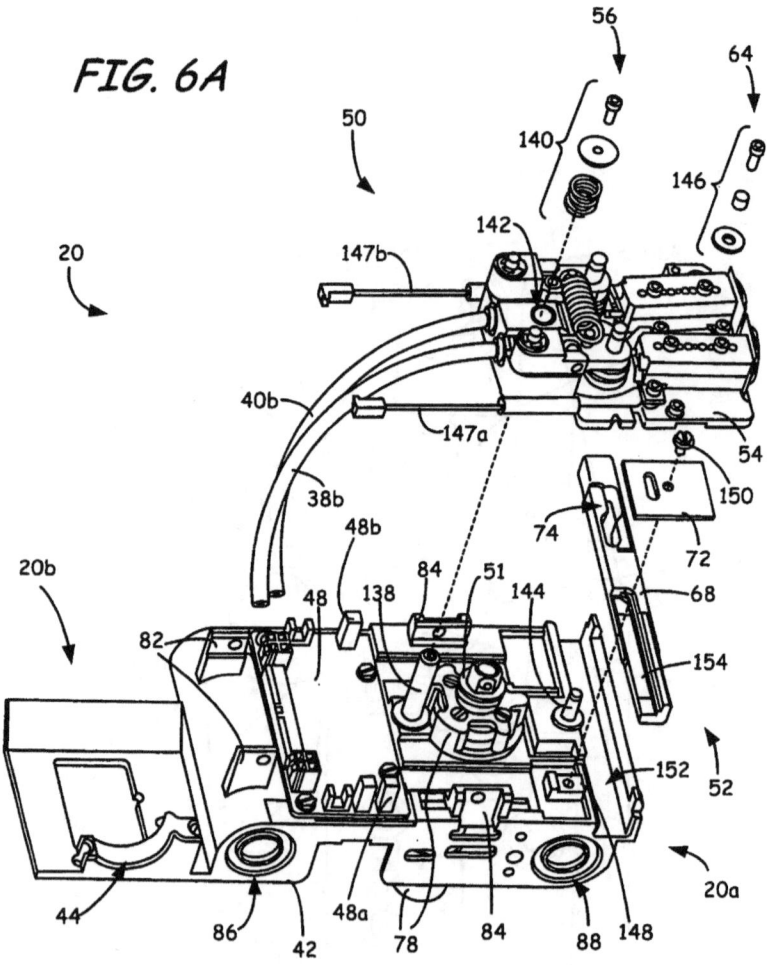

FIG. 6A

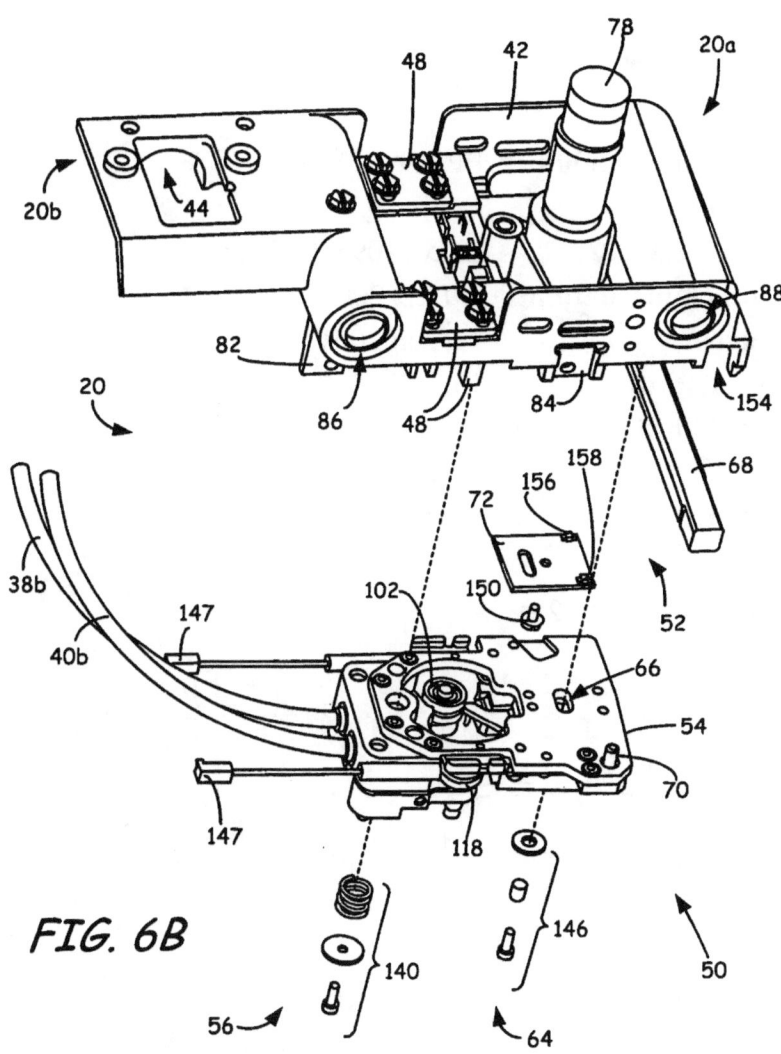

FIG. 6B

Filament Kassette

Veröffentlichungsnummer	EP1299217 B1
Publikationstyp	Erteilung
Anmeldenummer	EP20010952987
Veröffentlichungsdatum	1. Dez. 2004
Eingetragen	12. Juli 2001
Prioritätsdatum	13. Juli 2000
Auch veröffentlicht unter	CN1216726C, 18 weitere »
Erfinder	Steve Brose, 8 weitere »
Antragsteller	Stratasys Inc.

Zeichnungen (11 von 20)

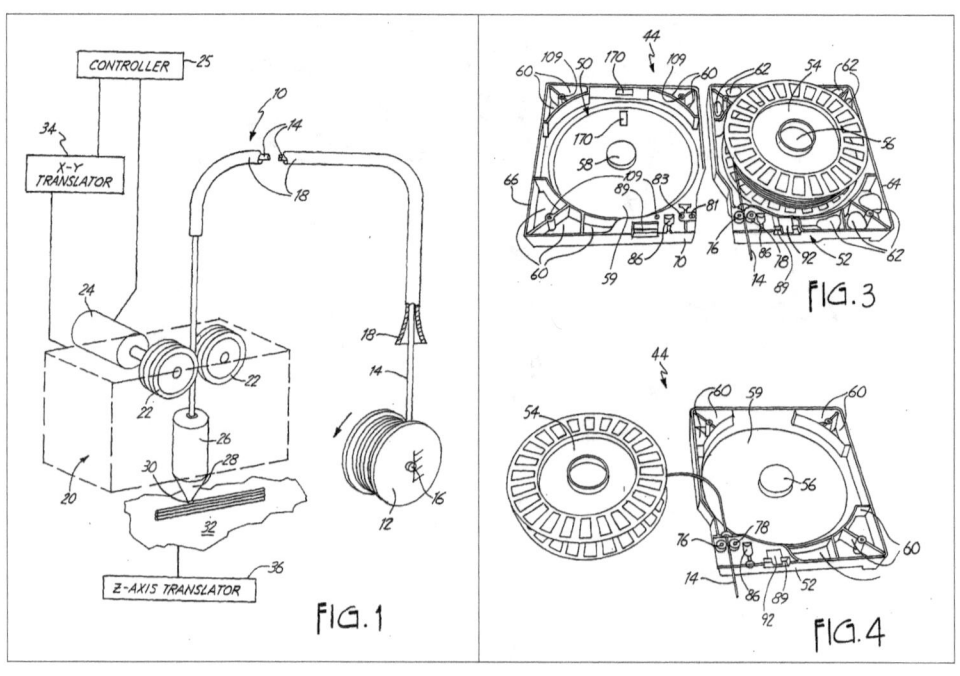

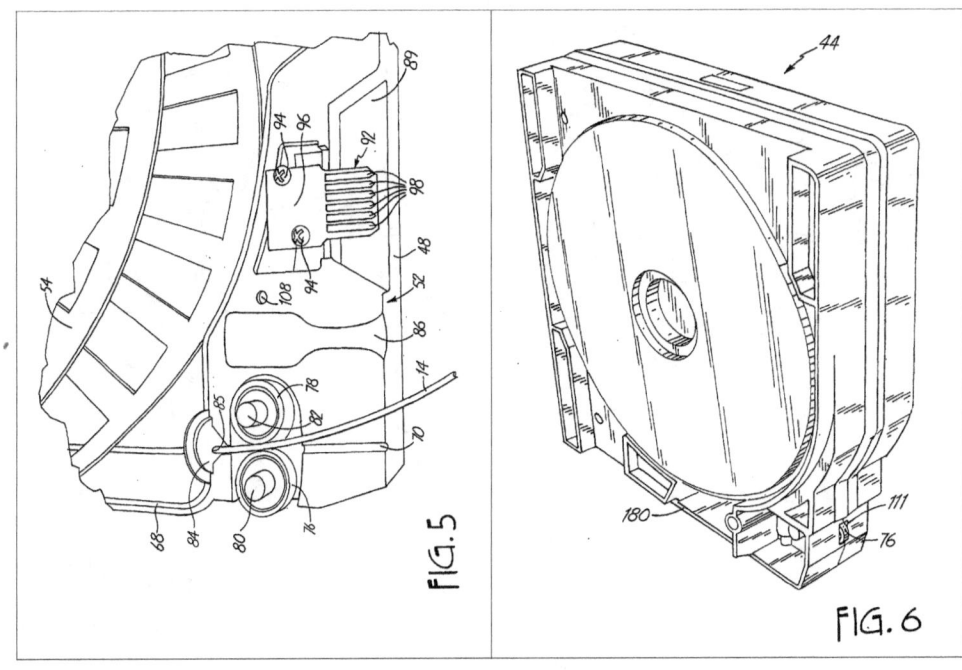

FIG. 5

FIG. 6

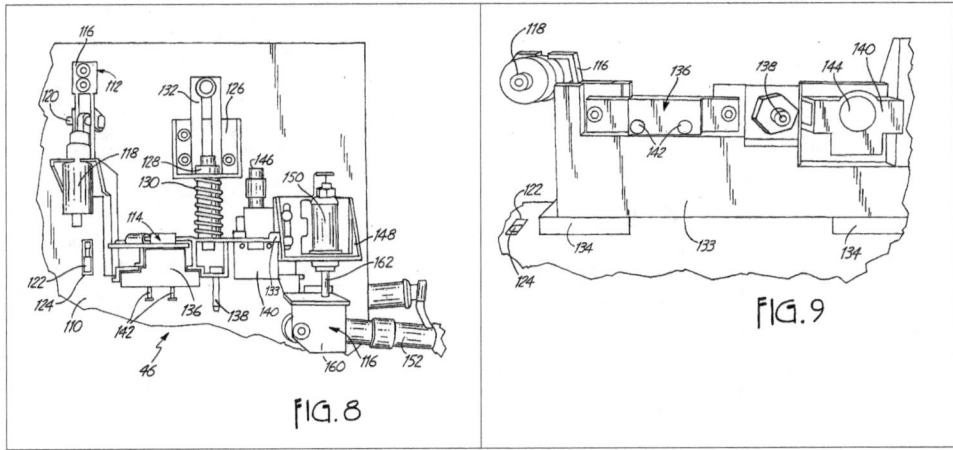

FIG. 8

FIG. 9

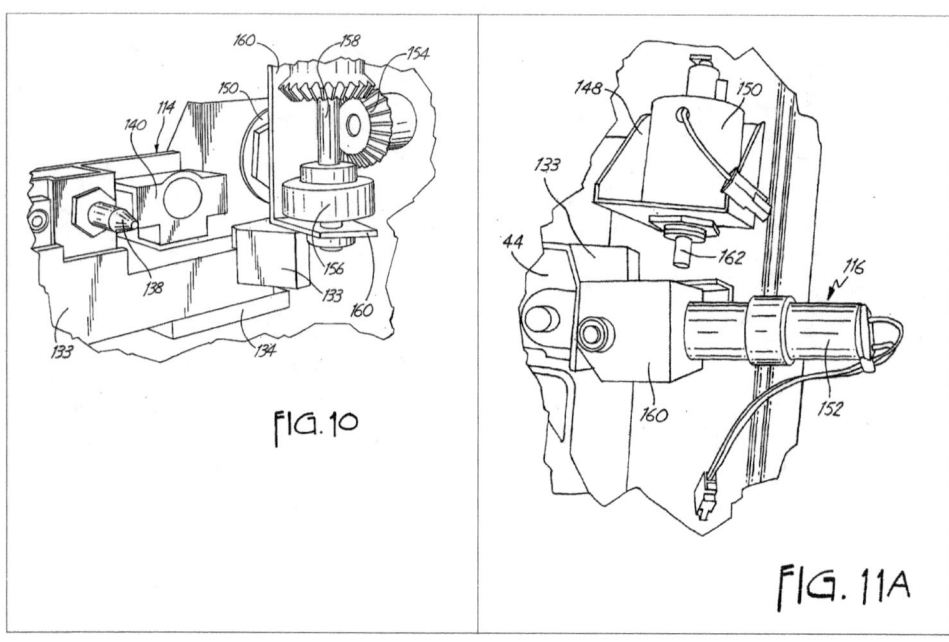

FIG. 10

FIG. 11A

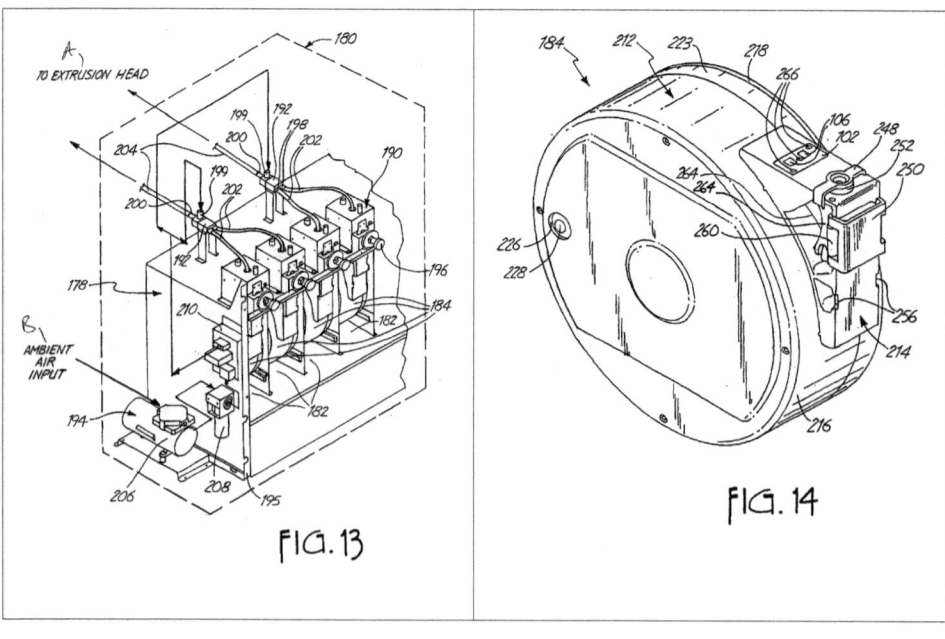

TO EXTRUSION HEAD

AMBIENT
AIR
INPUT

FIG. 13

FIG. 14

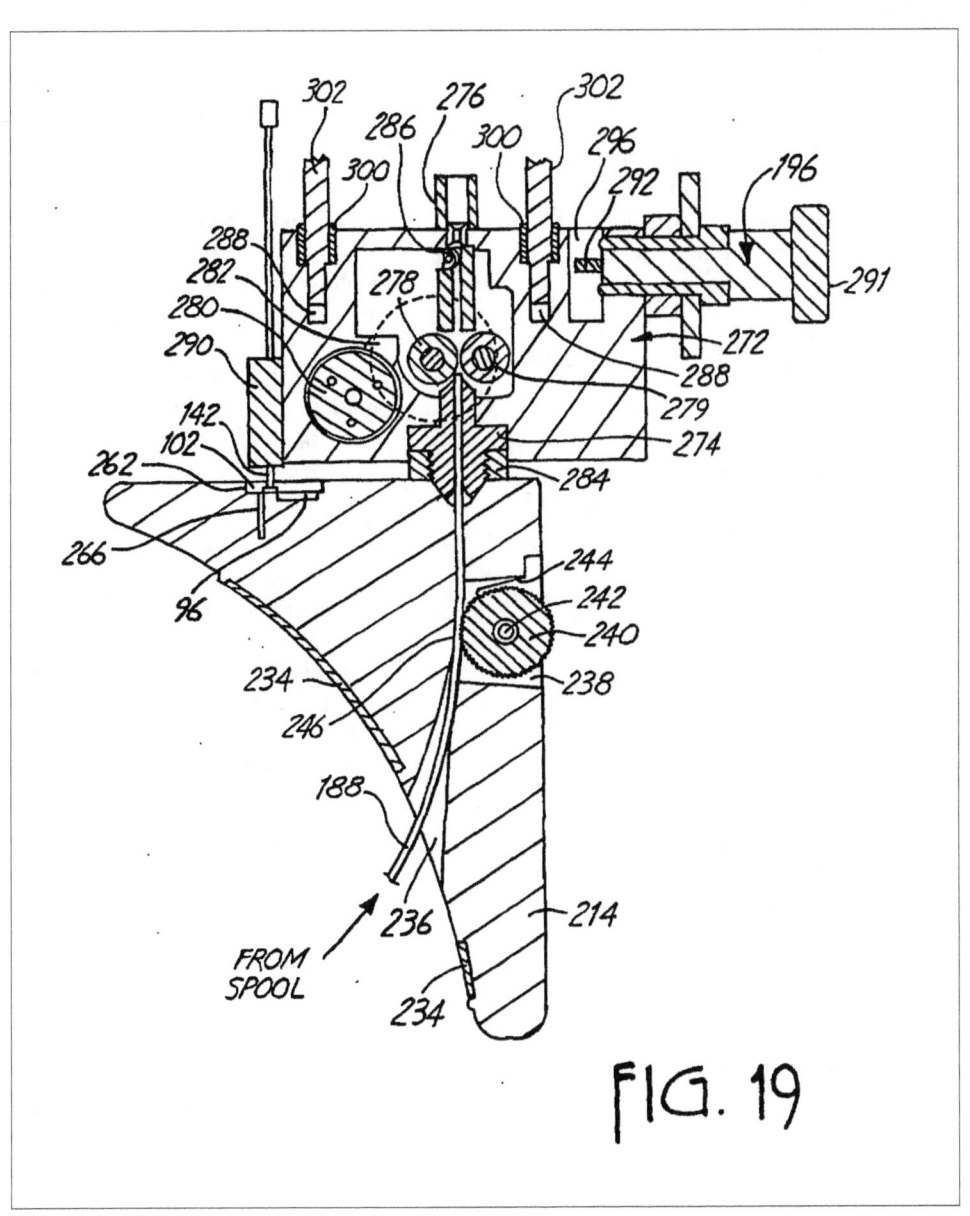

FIG. 19

ANSPRÜCHE

1. Filamentkassette (44, 184), die folgendes umfasst:

2. eine Kammer (59) mit einer drehbaren Spule (54, 186) auf der Modellierfilament (14, 188) aufgewickelt ist, die durch Erhitzen fließfähig gemacht wird;

3. einen Filamentpfad (70, 236), der von der Kammer (59) zu einer Austrittsöffnung (72, 238) führt und,

4. eine Vorrichtung (116, 278, 279, 280, 282) zum Vorwärtsbewegen eines Strangs des Filaments (14, 188) von der Spule (54, 186) entlang des Filamentpfades (70, 236),

5. dadurch gekennzeichnet, dass:

6. die besagte Kammer (59) einen Vorrat an Trockenmittel (62) enthält und,

7. die Kammer (59) luftdicht ist;

8. wobei die Filamentkassette (44, 184) zum Heranführen von Modellierfilament an eine dreidimensionale Modelliermaschine (40, 180) genutzt wird.

9. Filamentkassette nach Anspruch 1, wobei die Mittel zum Vorwärtsbewegen Folgendes umfassen:

10. ein Paar Rollen (76, 78), die einander gegenüberliegend entlang des Filamentpfades angebracht sind, so dass sie den Filamentstrang zwischen sich greifen.

11. Filamentkassette nach Anspruch 2, wobei jede Rolle aus dem erwähnten Rollenpaar passiv ist, und eine Rolle aus dem Paar eine Begleitrolle (76) ist, die für den Zugriff einer äußeren Antriebskraft zugänglich ist.

12. Filamentkassette nach Anspruch 3, wobei die Begleitrolle (76) eine rechtwinklig zum Filamentpfad schwimmende Drehachse aufweist, so dass sich die Begleitrolle vom Filamentpfad bei Fehlen einer von außen einwirkenden Kraft wegbewegt, was eine Druckentlastung auf einen Filamentstrang im Filamentpfad bewirkt.

13. Filamentkassette nach Anspruch 1, wobei die Vorrichtung zum Vorwärtsbewegen eine Rändelrolle (240) aufweist, die so gegenüber einer Wand des Filamentpfades angebracht ist, dass der Filamentstrang zwischen den Rollen gegriffen wird.

14. Filamentkassette nach Anspruch 5, wobei die Rändelrolle (240) so angebracht ist, dass sie für eine äußere Antriebskraft zugänglich ist.

15. Filamentkassette nach Anspruch 1, wobei die Vorrichtung zum Vorwärtsbewegen einen erhobenen Umriss in einer Wand des Filamentpfades umfasst, über dem ein Filamentstrang positioniert ist, wobei der erhobene Umriss derart zugänglich ist, dass sich eine äußere Antriebskraft auf den Filamentstrang ausüben lässt.

16. Filamentkassette nach Anspruch 7, wobei der erhobene Umriss durch die Fläche einer

Riemenspannrolle definiert wird.

17. Filamentkassette nach Anspruch 1, wobei eine Rückhalterung (84) den Filamentstrang im Filamentpfad (70) positioniert, bei gleichzeitiger Abdichtung der Luftströmung entlang des Filamentpfades (70).

18. Filamentkassette nach Anspruch 1, wobei der Zugang zur Vorrichtung zum Vorwärtsbewegen durch eine Tür (250) erfolgt, die eine kompressible Dichtung (258) auf einer inneren Fläche aufweist, um das Einströmen von Luft in die Kammer zu verhindern.

19. Filamentkassette nach Anspruch 1, wobei die Kammer und das aufgewickelte Filament getrocknet werden, bis der Wassergehalt weniger als 700 ppm beträgt.

20. Filamentkassette nach Anspruch 1, wobei eine herausnehmbare Dichtung (248) das Einströmen von Luft durch die Austrittsöffnung verhindert.

21. Filamentkassette nach Anspruch 1, die darüber hinaus Folgendes umfasst:

22. einen derart auf der Kassette aufgebrachten elektronisch ablesbaren und

23. beschreibbaren Datenspeicher (96), dass er einer äußeren Steuerung zugänglich ist, die Information über das Filament enthält.

24. Methode zum Zuführen von Modellierfilament an eine dreidimensionale Modelliermaschine, wobei diese

Methode die folgenden Schritte umfasst:

25. Einführen einer Kassette (44, 184) nach einem der Ansprüche 1 bis 13 in eine Aufnahmevorrichtung (42, 182) zum Beladen der Modelliermaschine (44, 184);

26. Einfädeln eines Filamentstrangs (14, 188) in den Filamentpfad (70, 236) der Kassette (44, 184) und,

27. Vorziehen des Filamentstrangs (14, 188) aus der Austrittsöffnung (72, 238) der Kassette (44, 184) und Einfädeln des Strangs in eine entsprechende Führung (140, 274) der Modelliermaschine.

28. Methode nach Anspruch 14, die darüber hinaus einen Schritt umfasst, bei dem das Strömen von Luft in die Kammer während des Ziehens von Filament (188) aus der Kassette (184) verhindert wird.

29. Methode nach einem der Ansprüche 14 oder 15, die darüber hinaus die folgenden Schritte umfasst:

30. Feststellen, dass das noch in der Kassette (44, 184) verbleibende Filament (14, 188) eine vorgegebene Mindestlänge erreicht hat und,

31. automatisches Rückführen des Filamentstrangs (14, 188) aus der Führung (140, 274), als Reaktion auf die Feststellung, dass die Mindestlänge erreicht wurde, so dass die Kassette (44, 184) entfernt und ausgetauscht werden kann.

32. Methode zum Zusammensetzen

der Filamentkassette gemäß einem der Ansprüche 1 bis 13, welche die folgenden Schritte umfasst:

33. Laden der Spule (54, 186) mit aufgewickeltem Filament (14, 188) in die Kammer (59) und,

34. Versiegeln der Kassette nach dem Beladen mit dem Filament (14, 188), so dass die Kammer (59) luftdicht wird.

35. Methode nach Anspruch 17, die darüber hinaus den folgenden Schritt umfasst:

36. Heizen der Filamentkassette (44, 184) in einem Ofen, bis der Wassergehalt der Kammer weniger als 700 ppm beträgt, bevor der Versiegelungsschritt erfolgt.

37. Filamentkassette nach Anspruch 1, die darüber hinaus Folgendes umfasst:

38. einen Hohlraum (88) zum Ausrichten der Kassette nach einer Vorrichtung zum Aufnehmen der Kassette (46) der Modelliermaschine.

Drucken in Graustufen oder Farbverläufen

Veröffentlichungsnummer	US20130095302 A1
Publikationstyp	Anmeldung
Anmeldenummer	US 13/478,233
Veröffentlichungsdatum	18. Apr. 2013
Eingetragen	23. Mai 2012
Prioritätsdatum	14. Okt. 2011
Erfinder	Nathaniel B. Pettis, Adam G. Mayer,Anthony James Buser
Ursprünglich Bevollmächtigter	Nathaniel B. Pettis, Adam G. Mayer,Anthony James Buser

Klassifizierungen (9), Legal Events (2)

Kommentar:

Die angemeldete Erfindung beschreibt ein Verfahren, bei des durch die Verwendung mindestens zweier verschiedener Filamente ein Objekt zu generieren, das in Graustufen oder in farbigen Verläufen entsteht, also somit ein echter dreidimensionaler Farbdruck. Die Intelligenz der Lösung liegt hier in der Software. Sie berechnet aus dem räumlichen Aufbau (Wandstärke usw.) dem Brechungsindex des Materials und den Farbwerten des Filaments eine räumliche Anordnung der Voxels die somit additiv zu dem gewünschten Effekt führen. Ein graues Objekt ist also aus mehr oder weniger durchsichtigem Material mit einem schwarzen Kern aufgebaut. Je nach Stärke des Kerns variiert die Graustufe, Für die additive Farbmischung gilt dasselbe. Auch unterschiedliche haptische Oberflächen (Texturen) will man so erzeugen.

Der Ansatz klingt interessant, denn hier wird die räumliche Eigenschaft des Objekts tatsächlich ausgenutzt, ob aber die relativ großen Schmelzpunkte einer FDM Düse fein genug ausgelegt werden können um diese optische Täuschung glaubhaft zu machen, muss sich noch beweisen.

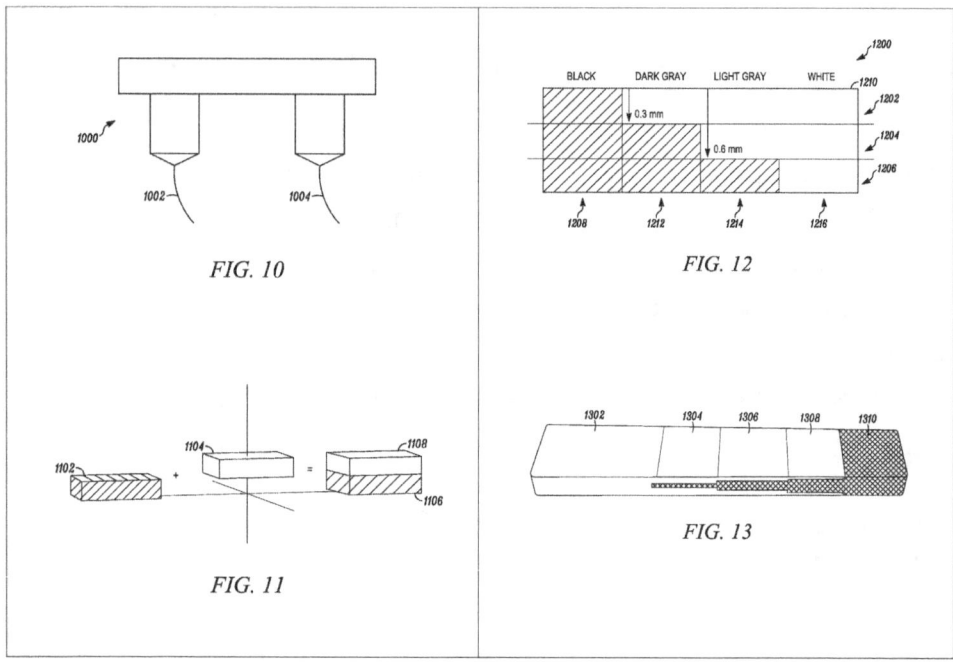

FIG. 10

FIG. 12

FIG. 11

FIG. 13

Kommentar 2:

Der Anmelder hätte, nach Auffassung des Autors, besser zwei einzelne Patentanmeldungen verfasst. Eine für die Farbmischung und eine für die Gestaltung der Textur. Es geht aus der Anmeldung (Summary) nicht klar hervor wie diese beiden Funktionen zusammen hängen, teilweise geht die Patentschrift hier auch durcheinander.

Durch die Nennung zweier Verfahren macht sich eine Patentanmeldung jedoch leichter angreifbar, da ein Wettbewerber o.ä. Eine Vielzahl von Ansätzen findet, bei denen er Ansprüche angreifen kann, die evtl. zurück genommen werden müssen.

Die Erfindung ist Stand 2013 bislang auch nur angemeldet, die Prüfung also noch nicht abgeschlossen.

Auch ist der originale Titel der Publikation „*Grayscale rendering in 3d printing*" etwas irreführend. Dies scheint auch der Anmelder Makerbot Inc. so zu sehen, denn er erklärt am Ende des Dokuments:

Accordingly, the term grayscale as used herein is intended to describe any value or array of values that specify portions of an image according to a controllable or selectable scale of values, and does not imply any specific color or translucence of build materials that are overlapped according to a "grayscale" image of such values

Warum aber eine Erfindung so nenne, wenn man es nicht so meint? Der Begriff Greyscale=Graustufen ist weithin geläufig und steht für eine bestimmte Art der Darstellung von weißen und schwarzen Bildpunkten. Warum Graustufen nun auf einmal für haptische Texturen und Farbverläufe herhalten sollen bleibt unklar. Die Gefahr diesen relativ alten Begriff zu verwenden, erhöht weiter das Risiko, dass die Ingenieure bei der Recherche zur Prüfung andere Patente aus dem 2D Druck Bereich finden, die einer Erteilung entgegen stehen. Das Patent, der wirtschaftlich äußerst erfolgreichen Firma Makerbot ist ein weiteres Beispiel für die Unsitte der dynamischen Branche, in Ermangelung intensiver Recherche oder Überlegung, Dinge den falschen Namen zu geben.

Glätten einer Oberfläche nach dem Druck

Veröffentlichungsnummer	EP1501669 B1
Publikationstyp	Erteilung
Anmeldenummer	EP20030716969
Veröffentlichungsdatum	24. Nov. 2010
Eingetragen	4. Apr. 2003
Prioritätsdatum	17. Apr. 2002
Auch veröffentlicht unter	CA2482848A1, 9 weitere »
Erfinder	JR. William R. c/o STRATASYS INC. PRIEDEMAN, 1 weitere »
Antragsteller	Stratasys, Inc.

Zeichnungen

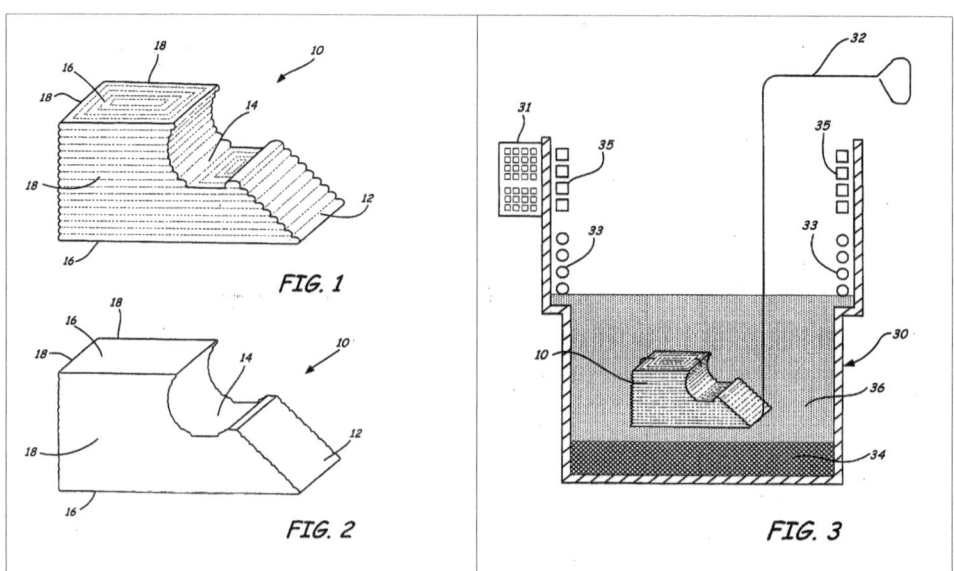

FIG. 1

FIG. 2

FIG. 3

ANSPRÜCHE

1. Verfahren zum Herstellen eines dreidimensionalen Objekts, das die folgenden Schritte umfasst:

2. Bereitstellen eines Objekts (10), das aus polymerem oder Wachsmodellier-Material mittels einer Layered Manufacturing Rapid Prototyping Technik (Schichtherstellung mit schnellem Prototypenbau) hergestellt wird, wobei das hergestellte Objekt eine Objektoberfläche (12, 14, 16,

18) aufweist, die aus dem Modelliermaterial ausgebildet ist, und wobei wenigstens ein Teil der Objektoberfläche aufgrund der "Layered Manufacturing Rapid Prototyping" Technik einen Oberflächeneffekt aufweist;

3. Aussetzen des Objekts mit Dämpfen eines Lösungsmittels (34), das das Modelliermaterial auf der Objektoberfläche vorübergehend aufweicht; und dadurch gekennzeichnet, dass es des Weiteren den folgenden Schritt umfasst: Wiederaufschmelzen des aufgeweichten Modelliermaterials, um den Oberflächeneffekt zu reduzieren.

4. Verfahren nach Anspruch 1, wobei die Schichtherstellungstechnik Fused Deposition Modeling (Schmelzschichtung) ist.

5. Verfahren nach Anspruch 1, wobei das Modelliermaterial ein thermoplastischer Kunststoff ist.

6. Verfahren nach Anspruch 3, wobei der thermoplastische Kunststoff wenigstens ca. 50 Gewichtsprozent eines amorphen Thermoplasten ausgewählt aus der Gruppe bestehend aus ABS, Polycarbonat, Polyphenylsulfon, Polysulfon, Polystyrol, Polyphenylenether, amorphe Polyamide, Acryle, Poly(2-Ethyl-2-Oxazolin) und Mischungen davon aufweist.

7. Verfahren nach Anspruch 1, wobei das Lösungsmittel ausgewählt wird aus der Gruppe bestehend aus Methylenchlorid, einer n-Propylbromid-Lösung, Perchlorethylen, Trichlorethylen und einer Hydrofluorcarbonat-Lösung.

8. Verfahren nach Anspruch 1, wobei das Modelliermaterial ausgewählt wird aus der Gruppe bestehend aus thermoplastischen Kunststoffen, in einem polymeren Bindemittel verteilten Rohmetallen, in einem polymeren Bindemittel verteilte Rohkeramik, und Strahlwachs.

9. Verfahren nach Anspruch 6, wobei das Modelliermaterial glasgefülltes Nylon ist.

10. Verfahren nach Anspruch 1, das des Weiteren den folgenden Schritt umfasst:

11. Auswählen der Zeitdauer, während der das Objekt den Lösungsmitteldämpfen auszusetzen ist, in Abhängigkeit von der Konzentration der Lösungsmitteldämpfe, vor dem Schritt des Aussetzens.

12. Verfahren nach Anspruch 8, das des Weiteren den folgenden Schritt umfasst:

13. Verringern der Konzentration der Lösungsmitteldämpfe, so dass sich die gewählte Zeitdauer des Aussetzens erhöht.

14. Verfahren nach Anspruch 1, das des Weiteren den folgenden Schritt umfasst:

15. Abdecken ausgewählter Abschnitte der Objektoberfläche mit einer Substanz, die das Glätten der ausgewählten Abschnitte verhindert, vor dem Schritt des Aussetzens des Objekts mit den Dämpfen des Lösungsmittels.

16. Verfahren nach Anspruch 10, wobei die Abdeckungssubstanz mittels eines automatischen Verfahrens aufgebracht wird.

17. Verfahren nach Anspruch 11, wobei das automatische Verfahren ein Sprühverfahren ist.

18. Verfahren nach Anspruch 11, wobei das automatische Verfahren ein Fused Deposition Modeling (Schmelzschichtungs-) Verfahren ist.

19. Verfahren nach Anspruch 11, das des Weiteren den folgenden Schritt umfasst:

20. Identifizieren der ausgewählten Abschnitte der Objektoberfläche zum Abdecken gemäß ihrer Geometrie.

21. Verfahren nach Anspruch 14, das des Weiteren den folgenden Schritt umfasst:

22. Identifizieren der ausgewählten Abschnitte der Objektoberfläche zum Abdecken gemäß ihrer Krümmungsradien.

23. Verfahren nach Anspruch 11, das des Weiteren den folgenden Schritt umfasst:

24. Identifizieren der ausgewählten Abschnitte der Objektoberfläche mittels eines Software-Algorithmus, der eine digitale Darstellung des zu schützenden Oberflächenbereichs erzeugt.

25. Verfahren nach Anspruch 16, wobei digitale Daten, die den zu schützenden Oberflächenbereich identifizieren, in einer .stl-Datei gespeichert werden.

26. Verfahren nach Anspruch 1, das des Weiteren den folgenden

Schritt umfasst:

27. Erzeugen einer digitalen Maske ausgewählter Abschnitte der Objektoberfläche, für die keine Glättung gewünscht wird, mittels einer haptischen Eingabeschnittstelle.

28. Verfahren nach Anspruch 1, wobei der Herstellungsschritt das Vordeformieren bestimmter Objektmerkmale (44, 46) umfasst, so dass diese Merkmale nach dem Schritt des Aussetzens eine gewünschte Geometrie (42) erhalten.

29. Verfahren nach Anspruch 19, das des Weiteren die folgenden Schritte umfasst:

30. Bereitstellen einer Anfangs-Objektdarstellung in einem digitalen Format, wobei die Anfangs-Objektdarstellung eine Oberflächengeometrie hat; und

31. Modifizieren der Anfangs-Objektdarstellung, um bestimmte Merkmale (44, 46) der Oberflächengeometrie vorzudeformieren, wodurch eine modifizierte Objektdarstellung entsteht;

32. wobei das in dem Herstellungsschritt hergestellte Objekt eine Geometrie hat, die gemäß der modifizierten Objektdarstellung definiert wird; und

33. wobei die nach dem Schritt des Aussetzens erhaltene gewünschte Geometrie annähernd mit der der Anfangs-Objektdarstellung übereinstimmt.

34. Verfahren nach Anspruch 1, wobei der Oberflächeneffekt

ausgewählt wird aus der Gruppe bestehend aus einem Treppeneffekt, einem Riefeneffekt und einer Kombination davon.

35. Verfahren nach Anspruch 21, wobei das Modelliermaterial ein thermoplastischer Kunststoff ist.

36. Verfahren nach Anspruch 22, wobei der thermoplastische Kunststoff wenigstens ca. 50 Gewichtsprozent eines amorphen Thermoplasten ausgewählt aus der Gruppe bestehend aus ABS, Polycarbonat, Polyphenylsulfon, Polysulfon, Polystyrol, Polyphenylenether, amorphes Polyamid, Methylmethacrylat, Poly(2-Ethyl-2-Oxazolin) und Mischungen davon aufweist.

37. Verfahren nach Anspruch 23, wobei das Lösungsmittel ausgewählt wird aus der Gruppe bestehend aus Methylenchlorid, einer n-Propylbromid-Lösung, Perchlorethylen, Trichlorethylen und einer Hydrofluorcarbonat-Lösung.

38. Verfahren nach Anspruch 21, wobei das Modelliermaterial ausgewählt wird aus der Gruppe bestehend aus thermoplastischen Kunststoffen, in einem polymeren Bindemittel verteilten Rohmetallen, in einem polymeren Bindemittel verteilter Rohkeramik, und Strahlwachs.

39. Verfahren nach Anspruch 25, wobei das Modelliermaterial glasgefülltes Nylon ist.

40. Verfahren nach Anspruch 21, das des Weiteren den folgenden Schritt umfasst:

41. Abdecken ausgewählter Abschnitte der Objektoberfläche mit einer Substanz, die das Aufweichen der ausgewählten Abschnitte verhindert, vor dem Schritt des Umschmelzens der Oberfläche.

42. Verfahren nach Anspruch 27, wobei der Abdeckungsstoff mittels eines automatischen Verfahrens aufgebracht wird.

43. Verfahren nach Anspruch 28, wobei das automatische Verfahren ein Sprühverfahren ist.

44. Verfahren nach Anspruch 28, wobei das automatische Verfahren ein Fused Deposition Modeling (Schmelzschichtungs-) Verfahren ist.

45. Verfahren nach Anspruch 28, das des Weiteren den folgenden Schritt umfasst:

46. Identifizieren der ausgewählten Abschnitte der Objektoberfläche zum Abdecken gemäß ihrer Geometrie.

47. Verfahren nach Anspruch 31, das des Weiteren den folgenden Schritt umfasst:

48. Identifizieren der ausgewählten Abschnitte der Objektoberfläche zum Abdecken gemäß ihrer Krümmungsradien.

49. Verfahren nach Anspruch 28, das des Weiteren den folgenden Schritt umfasst:

50. Identifizieren der ausgewählten Abschnitte der Objektoberfläche mittels eines Software-Algorithmus, der eine digitale Darstellung des zu schützenden Oberflächenbereichs erzeugt.

51. Verfahren nach Anspruch 33, wobei digitale Daten, die den zu schützenden Oberflächenbereich identifizieren, in einer .stl-Datei gespeichert werden.

52. Verfahren nach Anspruch 28, das des Weiteren den folgenden Schritt umfasst:

53. Identifizieren der ausgewählten Abschnitte der Objektoberfläche zum Abdecken mittels einer haptischen Eingabeschnittstelle.

54. (..)

55. Modifizieren der Anfangs-Objektdarstellung, um bestimmte Merkmale (44, 46) der Oberflächengeometrie vorzudeformieren, wodurch eine modifizierte Objektdarstellung erzeugt wird; und

56. Herstellen des Objekts gemäß der modifizierten Objektdarstellung aus dem Modelliermaterial mittels der Layered Manufacturing Rapid Prototyping Technik;

57. wobei das Umschmelzen des aufgeweichten Modelliermaterials zum Verringern des Oberflächeneffekts ein fertiges Objekt erzeugt, wobei das fertige Objekt eine Oberflächengeometrie hat, die annähernd mit der der Anfangs-Objektdarstellung übereinstimmt.

58. Verfahren nach Anspruch 36, das des Weiteren den folgenden Schritt umfasst:

59. Identifizieren von Merkmalen der Oberflächengeometrie zum Vorumformen gemäß ihren Krümmungsradien.

Simultanes Drucken von mehreren Lagen

Veröffentlichungsnummer	US8454345 B2
Publikationstyp	Erteilung
Anmeldenummer	US 12/836,582
Veröffentlichungsdatum	4. Juni 2013
Eingetragen	15. Juli 2010
Prioritätsdatum	16. Jan. 2003
Auch veröffentlicht unter	CA2513291A1, 109 weitere »
Erfinder	Kia Silverbrook
Ursprünglich Bevollmächtigter	Silverbrook Research Pty Ltd

Patentzitate (26), Klassifizierungen (20), Legal Events (2)

Kommentar:

Das ist Zukunftsmusik. Der Anmelder will gleichzeitig mehrere Lagen drucken. Hierzu wird ein feststehender Kopf mit multiplen Düsen vorgeschlagen und ein Transportband, dass eine Druckplattform unter diesem Kopf vorbei fährt. In der Anmeldung ist die Rede davon, einen LCD Fernseher zu drucken und zwar nicht nur das Gehäuse, sondern komplett mit allen Funktionsgruppen!

Ob die benötigten Komponenten, Werkstoffe und geforderten Güten tatsächlich durch Ejektion von Voxels erzeugt werden können, bleibt abzuwarten. Aber interessant ist das Konzept, den Ausdruck sowie die Farbe und Materialeigenschaften durch die Verwendung unzähliger Düsen zu beschleunigen oder erst zu ermöglichen.

Zeichnungen:

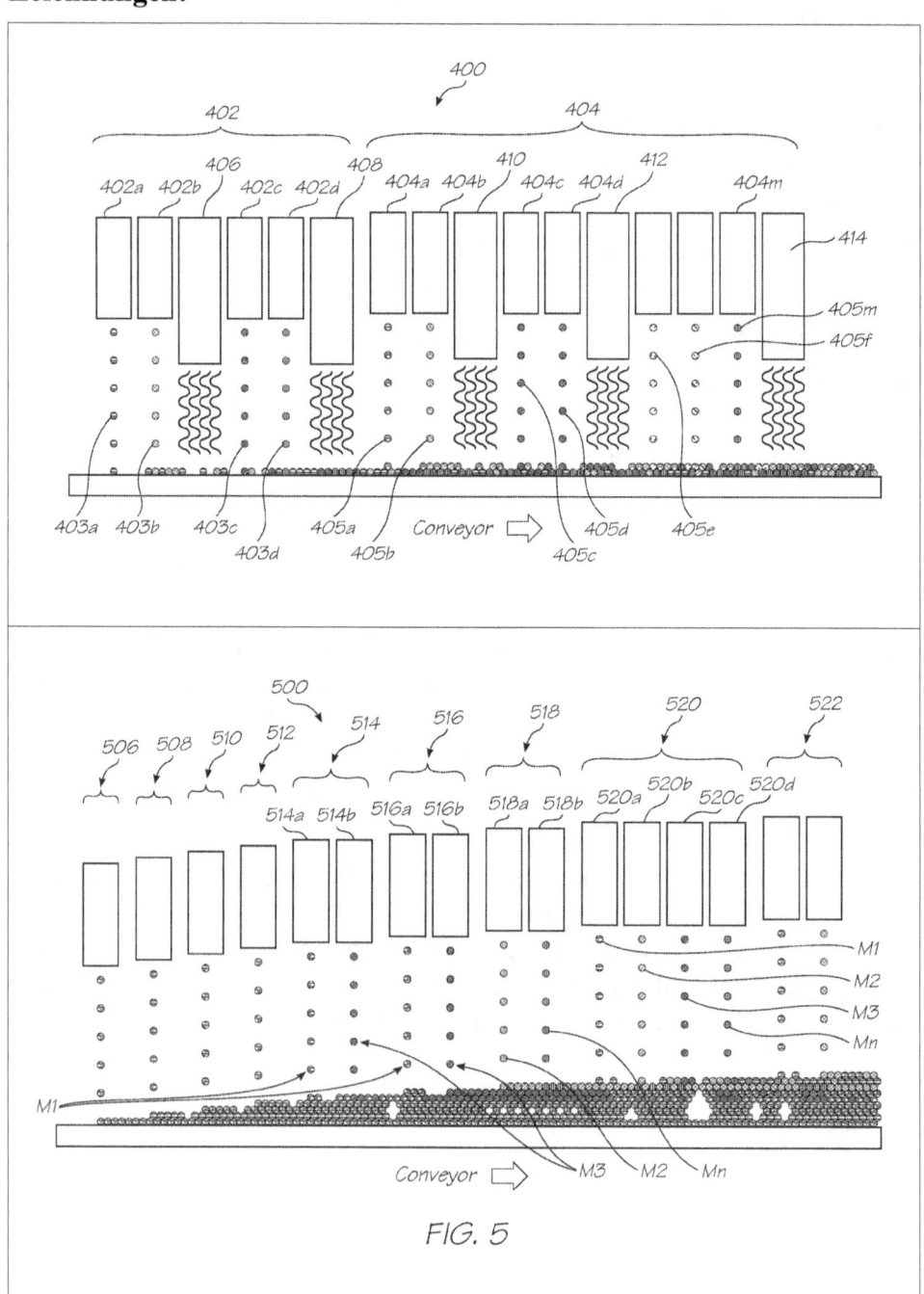

FIG. 5

Druckbare Support Struktur aus Silikon + Polymer

Veröffentlichungsnummer	US7534386 B2
Publikationstyp	Erteilung
Anmeldenummer	US 11/985,387
Veröffentlichungsdatum	19. Mai 2009
Eingetragen	15. Nov. 2007
Prioritätsdatum	20. Apr. 1999
Gebührenstatus	Bezahlt
Auch veröffentlicht unter	CN1666217A, 6 weitere »
Erfinder	William R. Priedeman, Jr.
Ursprünglich Bevollmächtigter	Stratasys, Inc.

Patentzitate (90), Klassifizierungen (51), Legal Events (3)

Externe Links: USPTO, USPTO-Zuordnung, Espacenet

Kommentar:

Der Erfinder beschreibt ein Verfahren, bei dem ein Dualextruder einen thermoplastistischen Werkstoff(A) und eine silikonbasierende Materialmischung(B) zum Drucken von Model und Stützstruktur verwendet. Hierbei gibt es einige Probleme zu lösen, denn die Stützstrukturen müssen ähnliche Temperaturen aushalten wie das eigentliche Modell, trotzdem sich später noch einfach lösen lassen. Die Patentschrift nennt detaillierte Zusammensetzungen, Hersteller und Bezeichnungen der verwendeten Chemikalien, sowie die Ergebnisse von Tests. Bei einer genannten Mischung wurde eine mögliche Druckdauer von maximal 20 Stunden festgestellt. Danach verfärbte sich das auf Silikon basierende Material und haftete verstärkt an dem Model an, so dass eine Trennung sehr aufwendig wurde.

Die Erkenntnisse aus diesen Versuchen, samt Angabe der verwendeten Stoffe, Arbeitstemperaturen etc. dürfte dem interessierten Leser hilfreich sein. Auf den Bildern wurde nur eine Kartusche mit dem Hauptmaterial gezeigt. Zum Verfahren gehören jedoch immer zwei Komponenten und somit auch zwei Kartuschen.

Eine erstaunliche Erkenntnis liefern die durchgeführten Versuche: Mit einem Filament aus einer Mischung von einem Polymer (beispielsweise Nylons, PEEK, PEAK oder ABS, u.a.) und ca. 10% Silikon, ließen sich Stützstrukturen(B) schaffen, die sich dann von dem eigentlichen Objektmaterial, für das ebenfalls ein Polymer verwendet wurde, leicht trennen ließ und **die Verstopfung der Düse verhinderte.** Die Erfinder sprechen davon, dass bei einem typischen Stratasys 3D Drucker eine Standzeit von ca. 3.5 Kg Material besteht, bevor die Düse aufgrund angesetzter Verunreinigungen erneuert werden muss, während es beider Verwendung von Polymer plus 10% Silikon, bis zu 20 Kg Filament Material verbraucht werden konnte, ohne, dass sich die Düse zusetzte. Diese Erkenntnis war zwar nicht Ziel der Versuche, wurde jedoch auch mitgeteilt.

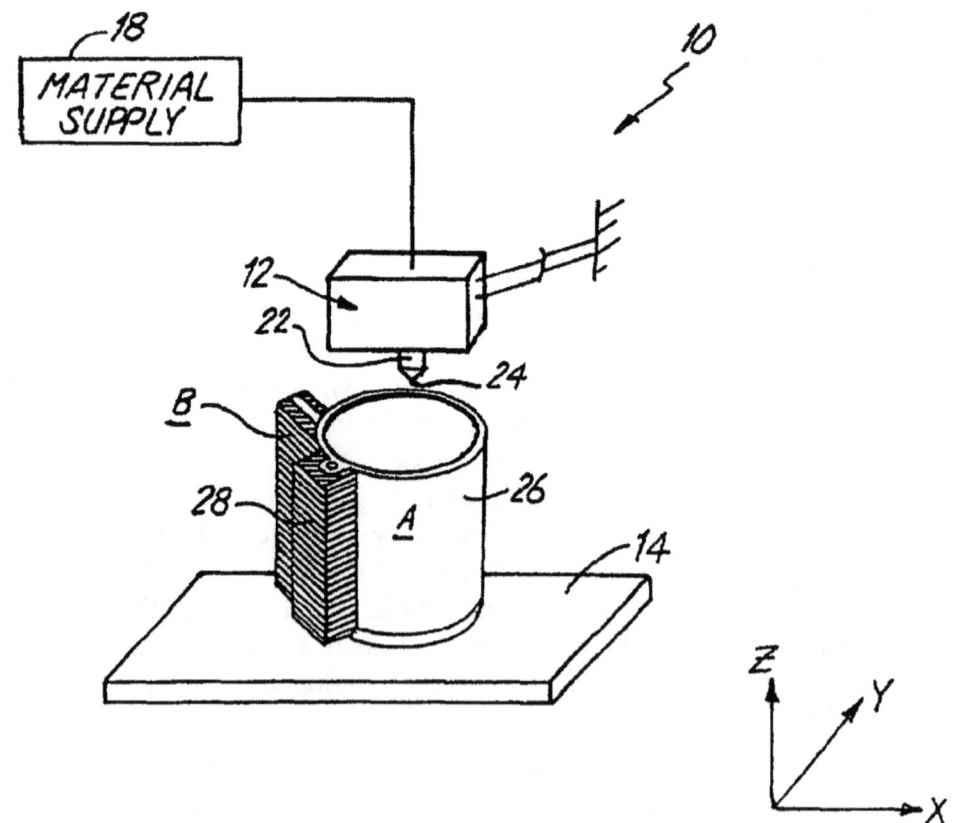

ZUSAMMENFASSUNG

A three-dimensional model and its support structure are built by fused deposition modeling techniques, wherein a thermoplastic material containing silicone is used to form the support structure and/or the model. The thermoplastic material containing silicone exhibits good thermal stability, and resists build-up in the nozzle of an extrusion head or jetting head of a three-dimensional modeling apparatus. The silicone contained in a support material acts as a release agent to facilitate removal of the support structure from the model after its completion.

SUMMARY

The present invention relates to a thermoplastic material for use in layered-deposition three-dimensional modeling. The thermoplastic material has a heat deflection temperature greater than about 220° C., a suitable melt flow for extrusion, and includes a base polymer and a silicone release agent constituting about 0.5 percent by weight to about 10 percent by weight of the thermoplastic material.

BRIEF DESCRIPTION OF THE DRAWINGS

FIG. 1 is a diagrammatic illustration of a model and a support structure therefor formed using layered extrusion techniques.

DETAILED DESCRIPTION

The present invention is described with reference to a deposition modeling system of the type shown in FIG. 1. FIG. 1 shows an extrusion apparatus 10 building a model 26 supported by a support structure 28 according to the present invention. The extrusion apparatus 10 includes an extrusion head 12, a material-receiving base 14 and a material supply 18. (..)

A modeling material A is dispensed to form the model 26, and a support material B is dispensed in coordination with the dispensing of modeling material A to form the support structure 28. For convenience, the extrusion apparatus 10 is shown with only one material supply 18. It should be understood, however, that in the practice of the present invention, the modeling material A and the support material B are provided to the extrusion apparatus 10 as separate feedstocks of material from separate material supplies. The extrusion apparatus 10 may then accommodate the dispensing of two different materials by: (1) providing two extrusion heads 12, one supplied with modeling material A and one supplied with support material B (..)

To properly support the model under construction, the support material B must bond to itself (self-laminate). The support materials B must form a weak, breakable bond to modeling material A (co-laminate), so that it can be separated from the completed model without causing damage to the model. Where the support structure is built up from the base, support material B must additionally bond to the base (..)

To produce a dimensionally accurate model, the modeling and support materials must exhibit little shrinkage upon cooling in the conditions of the build envelope. Any shrinkage of the support material B must match that of the modeling material A. A shrink differential in the materials would cause stresses and bond failures along the model/support structure joint. (..)

Testing of Materials:

The following are examples of material formulations which were tested for use as support materials in a very high-temperature modeling environment (i.e. build chamber temperature of 200° C. or greater). The material formulations were tested as support materials for a polyphenylsulfone modeling material.

74 Ausgewählte Patente

Specifically, in each case, the polyphenylsulfone modeling material is Radel™ R 5600 NT (available from BP Amoco). This polyphenylsulfone resin has a heat deflection temperature of 236° C., and a melt flow in the range of 20-30 gms/10 min. at 400° C. under a 1.2 kg load. Example 3 embodies the present invention, while Example 1 and 2 are comparative examples.

All of the materials tested met the rheology criteria discussed above. In each case, techniques conventional in polymer chemistry were used to compound the component materials. The exemplary materials were successfully formed into modeling filament of a very small diameter, on the order of 0.070 inches, and used in a filament-fed deposition modeling machine.

EXAMPLE 1

Models of various sizes were built in a build chamber having a temperature of about 200-225° C., using the polyphenylsulfone modeling material and a support material comprising a blend of polyphenylsulfone and amorphous polyamide. In some cases, the support material further included polysulfone. Weight percent ranges of the various component materials were between about 60 and 90 weight percent polyphenylsulfone, and between about 10 and 40 weight percent amorphous polyamide blend, or between about 60 and 90 weight percent polyphenylsulfone, between about 1 and 40 weight percent polysulfone and between about 10 and 40 weight percent amorphous polyamide blend. A particular exemplary resin tested is a blend of 50 weight percent Radel™ R 5600 NT polyphenylsulfone (available from BP Amoco), 25 weight percent Udel™ P 1710 NT 15 polysulfone (available from BP Amoco), and 25

weight percent EMS TR 70 amorphous polyamide (available from EMS-Chemie AG of Switzerland). This resin has a heat deflection temperature of 224° C. and a melt flow similar to that of the modeling material. The support material was extruded from a liquifier having a temperature of about 350° C. to form a support structure for a model built using the polyphenylsulfone resin.

The support material according to this example was satisfactory for models that took less than about 20 hours to build, but failed for models that had a longer build time. It was observed that the support material exhibited thermally instability after about 20 hours in the build chamber. The thermally instability manifested by the material becoming dark and eventually blackening, and becoming strongly adhered to the model. Desirably, a material will survive build times of up to about 200 hours, to permit the building of large and complex parts. Thus, while the support material of the present example was found satisfactory for supporting small parts, it is not suitable for more general high-temperature use.

EXAMPLE 2

Test models were built in a build chamber having a temperature of about 200-225° C., using the polyphenylsulfone modeling material and a support material which comprised various resins of polyethersulfone, polyphenylsulfone or polyetherimide (i.e., Ultem™). These materials exhibited favorable thermal stability, but could not be broken away from the model. The support material containing polyphenylsulfone adhered very strongly to the model. The support material containing polyetherimide adhered fairy strongly to the model, and the support material containing polyethersulfone, while exhibiting the least adherence to the model, adhered too strongly for suitable use.

EXAMPLE 3

Large and small polyphenylsulfone models were built in a build chamber having a temperature of about 200-225° C., using a support material comprising a polyethersulfone base polymer and a silicone release agent. For convenience, commercially available compounds were used to provide a "masterbatch" containing silicone, which was compounded with the base polymer. Various masterbatches were tested, which included polypropylene, linear low-density polyethylene, and high-impact polystyrene. Additionally, various silicones were tested, ranging in viscosity from about 60,000 centistokes (intermediate viscosity) to 50 million centistokes (very high viscosity). The very high viscosity silicones have a high molecular weight, while the lower viscosity silicones have a lower molecular weight.

It was found that intermediate viscosity silicone was a much better release agent than the very high viscosity silicone, and that the high-impact polystyrene masterbatch released more easily from the polyphenylsulfone modeling material than did the other masterbatches tested. In a preferred embodiment, the masterbatch contained about 75 weight percent of a high-impact polystyrene copolymer and about 25 weight percent of a 60,000 centistoke (cSt) viscosity silicone. In this embodiment, the support material comprised between about 90-95 weight percent polyethersulfone, between about 3-8 weight percent high-impact polystyrene, and between about 1-3 weight percent silicone. This composition was demonstrated using BASF, Ultrason E-1010 polyethersulfone and Dow-

Corning MB25-504 styrene butadiene copolymer containing hydroxy-terminated poly dimethyl siloxane (i.e. hydroxy-terminated silicone). This material was extruded from a liquifier having a temperature of about 420° C. to successfully form a support structure for a model built using the polyphenylsulfone resin. The support structure satisfactorily released from the model after its construction.

The support material of the present example exhibited a tensile strength of between 5000 psi and 12,000 psi, exhibited a shrinkage typical of amorphous polymers (less than 0.010 inch/inch), a melt flow in the range of about 5-30 gms/10 min. under a 1.2 kg load at a temperature of up to 450° C., and a heat deflection temperature of about 232° C.

Discussion of Results

It was demonstrated that adding a small amount of silicone to a base polymer weakened the bond between the base polymer and the modeling material, enabling use of the polymer to form a support structure that could be broken-away from the model. An intermediate viscosity silicone (on the order of about 104-105 centistokes) provided good release characteristics, although it is expected that a variety of silicones can be used to advantage in the present invention.

As the silicone release agent exhibited thermal resistance at temperatures of 225° C. for over 200 hours, the present invention is particularly useful in supporting models made from high-temperature thermoplastics in a very hot environment. Heretofore, there have been no known materials suitable for building a support structure by layered deposition modeling techniques in an environment hotter than about 180° C.

While the composition of the present invention was demonstrated using a polyethersulfone base polymer, the silicone release agent can be added to a variety of other base polymers to likewise lessen adhesion of the support structure to the model. A base polymer is selected based upon various physical, thermal and Theological properties demanded by the deposition modeling process. For high-temperature processes, silicone added to a polyphenylsulfone or polyetherimide base polymer will exhibit good thermal stability. Other potential base polymers for use in various build environments include polyphenylenes, polycarbonates, high-impact polystyrenes, polysulfones, polystyrenes, acrylics, amorphous polyamides, polyesters, nylons, PEEK, PEAK and ABS. Where adhesion between a base polymer and a modeling material is higher, a greater amount of silicone can be added. A suitable amount of silicone will weaken but not destroy the bond between the support structure and the model, providing adhesion sufficient to support the model under construction. It is expected that up to about 10 weight percent silicone may be desired in some cases.

While a high-impact polystyrene co-polymer was used in demonstrating the present invention, such co-polymer is but one example of a copolymer which may be included in the composition of the present invention. The high-impact polystyrene masterbatch was used as a matter of convenience in compounding the silicone with the base polymer. Those skilled in the art will recognize that various masterbatches may be used (e.g., one made with the base polymer of the support material), that other techniques for compounding may be used which do not require a masterbatch (e.g., liquid silicone could be added directly to the base polymer), and that various other co-polymers may be included in the thermoplastic composition, in various amounts, to satisfy processing demands of a given application.

An unexpected benefit of the thermoplastic material containing silicone is that this material resisted build-up in the nozzle of the extrusion head liquifier. This attribute of the material, though unintended, is highly desirable. Typically, the liquifier of an extrusion-based layered deposition modeling machine needs to be replaced after extrusion of only about 7 pounds of material, due to an unacceptable build up of material in the nozzle. Resistance to clogging of the material containing silicone was observed to surpass that of any materials heretofore known in the art. The nozzles of liquifiers used to extrude the thermoplastic material containing silicone extruded over 40 pounds of the material before needing replacement. Nozzle life was thus extended by over 400 percent. Hence, silicone was demonstrated to provide the thermoplastic with characteristics desirable for modeling materials as well as support materials.

Resistance to nozzle clogging was demonstrated with compositions that included as little as 0.75 weight percent silicone. For modeling materials, the amount of silicone in the material may thus be kept very small, between about 0.5 weight percent and 2 weight percent, to prolong the liquifier life without degrading the strength of the modeling material. As will be recognized by those skilled in the art, the higher viscosity silicone, which has a lesser release ability, may be beneficial as an additive to the modeling material. As will be further recognized by those skilled in the art, where silicone is contained in both the modeling and support material, a reduced amount of silicone in the support material may be preferred.

Also as will be recognized by those skilled in the art, the modeling material A and support material B may include inert and/or active filler materials. The fillers can provide enhanced material properties which may be desirable depending upon the intended use of the resulting model. For instance, fillers can provide RF shielding, conductivity, or radio opaque properties (useful for some medical applications). Fillers can alternatively degrade material properties, but this may be acceptable for some uses. For instance, an inexpensive filler can be added to the modeling material A or support material B to decrease the cost of these materials. Fillers can also change thermal characteristics of the materials, for instance a filler can increase the heat resistance of a material, and a filler can reduce material shrinkage upon thermal solidification. Exemplary fillers include glass fibers, carbon fibers, carbon black, glass microspheres, calcium carbonate, mica, talc, silica, alumina, silicon carbide, wollastonite, graphite, metals and salts.

Filler materials which will assist in removal of the support structure can also be used in the composition of the present invention. For instance, a filler material that swells when contacted by water or another solvent will tend to be useful in breaking down the support structure. A filler material that evolves gas when contacted by water or another solvent will likewise tend to be useful in breaking down the support structure.

Those skilled in the art will recognize that innumerable other additives may also be to modify material properties as desired for particular applications. For instance, the addition of a plasticizer will lower the heat resistance and melt flow of a thermoplastic material. The addition of dyes or pigments can be done to change color. An antioxidant can be added to slow down heat degradation of material in the extruder.

The modeling and support materials A and B of this foregoing examples may be molded into filament, rods, pellets or other shapes for use as a modeling feedstock, or it may be used as a liquid feedstock without prior solidification. Alternatively, the mixture may be solidified and then granulated.

It is noted that the modeling material A and support material B of the foregoing examples are moisture sensitive. It has been demonstrated that exposure of these materials to a humid environment will significantly degrade model quality, thus, maintaining dry conditions is important. In order for the materials of the present invention to build accurate, robust models by fused deposition techniques, the material must dried. Particularly suitable apparatus for building up three-dimensional objects using the high temperature, moisture-sensitive materials of the present invention are disclosed in pending U.S. application Ser. Nos. 09/804,401 and 10/018,673, which are incorporated by reference herein. The '673 application discloses a modeling machine having a high-temperature build chamber, and the '401 application discloses a moisture-sealed filament cassette and filament path for supplying moisture-sensitive modeling filament in a filament-fed deposition modeling machine.

For the modeling material A and support material B of the foregoing examples, an acceptable moisture content (i.e. a level at which model quality will not be impaired) is a level less than 700 parts per million (ppm) water content (as measured using the Karl Fischer method). The '401 application discloses techniques for drying the filament provided in the a filament cassette. One method for drying the material is to place a cassette containing the material in an oven under vacuum conditions at a suitable temperature (between 175-220° F. is typical) until the desired dryness is reached, at which time the cassette is sealed. The cassette may then be vacuum-sealed in a moisture-impermeable package, until its installation in a machine. An expected drying time is between 4-8 hours to reach less than 300 ppm water content. Another method is to dry the material by placing packets of desiccant in the cassette without use of the oven. It has been demonstrated that placing packets containing Tri-Sorb-molecular sieve and calcium oxide (CaO) desiccant formulations in the cassette and sealing the cassette in a moisture-impermeable package will dry the material to a water content level of less than 700 ppm, and will dry the material to the preferred range of 100-400 ppm. This desiccant-only drying method has advantages over the oven-drying method in it requires no special equipment, and is faster, cheaper and safer than oven drying. Suitable Tri-Sorb-molecular sieve desiccant formulations include the following: zeolite, NaA; zeolite, KA; zeolite, CaA; zeolite, NaX; and magnesium aluminosilicate.

The '401 application further discloses a filament delivery system and an active drying system which will preserve the dryness of the material when it is loaded in the modeling machine. The drying system creates an active moisture barrier along a filament path from the cassette to the extrusion head, and purges humid air from the modeling machine. The drying system continuously feeds dry air or other gas under pressure to the filament path, disallowing humid air from remaining in or entering the filament path, and is vented at or near the end of the filament path.

Volumetrische Messung des Filaments

Publication number	WO1997037810 A1
Publication type	Application
Application number	PCT/US1997/005590
Publication date	16 Oct 1997
Filing date	3 Apr 1997
Priority date	8 Apr 1996
Also published as	US6085957
Inventors	John S Batchelder, Robert L Zinniel
Applicant	Stratasys Inc

Patent Citations (7), Referenced by (6), Classifications (14), Legal Events (6)

External Links: Patentscope, Espacenet

Kommentar: Das Verfahren beschreibt verschiedene Methoden um kontinuierlich den Querschnitt des Filaments zu messen um daraus da geförderte Volumen zu errechnen und dieses Ergebnis in die Extrusion als Korrekturfaktor einfließen zu lassen. Auf diese Weise werden die gefertigten Modelle präziser und reproduzierbar, trotz schwankender Filamentdicke. Genannt werden rein mechanische Messverfahren, wie das Andrücken von federbelasteten Rollen, deren Auslenkung ausgelesen und umgewandelt wird, ebenso wie optische Messmethoden bei denen zwei im Winkel von 90° angeordnete Lichtquellen von CCD Sensoren ausgelesen werden und andere Verfahren.

Aufgrund der OCR Erfassung des Textes durch Google kann dieser Fehler enthalten.

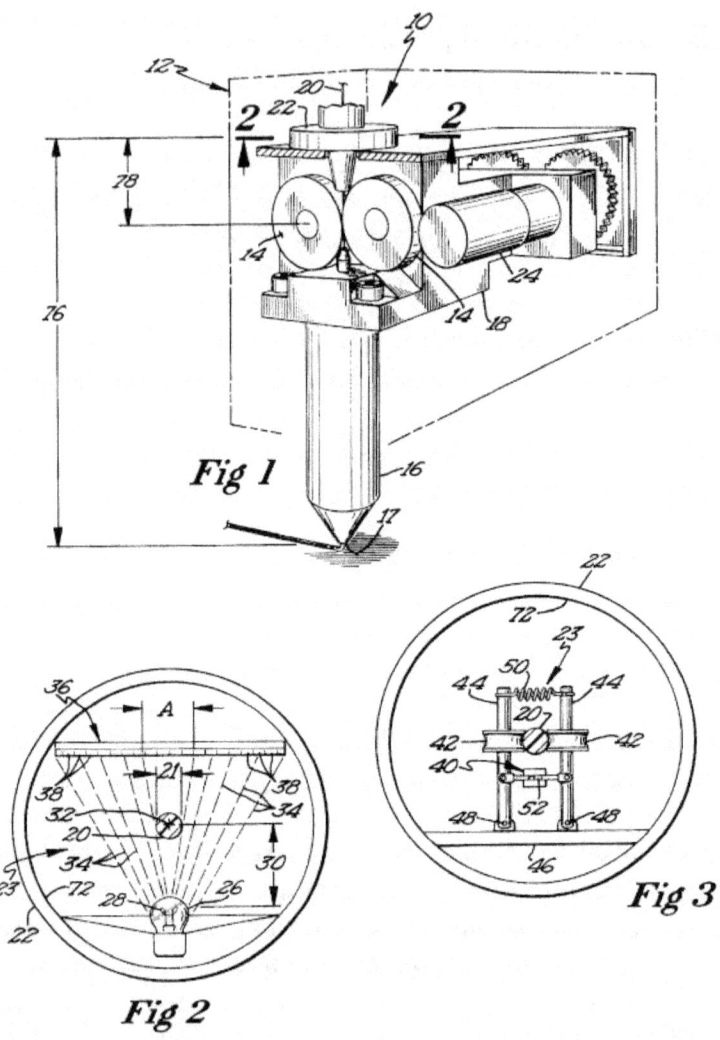

Fig 1

Fig 2

Fig 3

BRIEF SUMMARY OF THE INVENTION

It is an object of the present invention to provide a volumetric feed control system that will allow for an increased tolerance range in build element effective cross section, in other words, a loosening of the strictness of the tolerance requirements.

It is another object of the present invention to improve the constancy of the flow rate

of fluid delivered through the application tip of a three-dimensional modeling machine.

It is yet anotner object of the present invention to provide a feed control system suitable for installation on an existing three-dimensional modeler including the type disclosed in U.S.Patent No. 5, 121 , 329 to Crump, which is hereby incoφorated by reference, and which will require little modification of existing equipment. The present invention achieves these objects by providing a system for continuously measuring, computing, and monitoring the effective cross section of a build material element such as a filament as the element is fed to the melting unit or application tip of the modeling machine, and adjusting the speed at which the element is fed to the melting unit or the application tip to ensure a more constant flow rate of fluid at the application tip. Build material element is fed to a modeling machine by an advancement mechanism including a motor. The motor can be a stepper motor or a DC servo motor. In the present invention, the build element is advanced to the dispensing head by the advancement mechanism. The effective cross section of the element is determined through the use of sensing means and a central processing unit. Various configurations for determination of effective cross section may be used, for example infrared emitter and detector pairs, tungsten filament and infrared detector arrays, pinch rollers and linear variable differential transformers (LVDT), and capacitive measurement. There is usually a spatial gap between the sensing means and the application tip of the three-dimensional modeler. Therefore, a lag response exists in the system. A variance in filament cross section will not immediately affect the volume of filament material present at the application tip. To combat this delay, the distances between the sensing means and the application tip as well as between the sensing means and the center of the modeler advancement mechanism are provided to the central processing unit. The sensing means provide continuous signals to the central

processing unit which can then compute the effective cross section of the filament. The motor of the advancement mechanism is connected to the central processing unit to allow the central processing unit to know the speed of the motor at all times. Since the central processing unit knows the speed of the motor and receives measurements from the sensing means, the effective cross section of the element may be continuously computed. Necessary changes to motor speed and hence feed roller speed and the proper times to effect the changes in speed may therefore be controlled by the central processing unit to ensure that the lag response is continuously and properly compensated. The volumetric feed control apparatus of the present invention will allow the tolerance of flexible filament to be increased. In other words, the filament diameter accuracy level required by the present invention is less than that previously required by the prior art. The reason for this is the use of more accurate methods of providing a constant flow rate of filament to the modeling machine application tip. This relaxed tolerance requirement will considerably cut the manufacturing costs of filament, and therefore of models created by modeling machines.

Further, any existing system using a roller feed method of feeding a flexible filament to a modeler may easily use the present invention. Insertion of the sensing means into an existing system may be easily accomplished. The central processing unit may be located anywhere, provided that proper connections are made to the sensors and to the motor controlling the advancement mechanism. No changes other than the drive control of the motor, and the possible addition of an encoder f edback to the motor are necessary.

Typically, the build material element is configured in the shape of a cylindrical filament, normally wound on a spool or stored in a roll. However, the element may be supplied in a wide variety of configurations, including ribbons, tubes, extrusions of triangular, trapezoidal, or pentagonal shape, and the like. The build material element may also be supplied in discrete quantities, and need not be wound on a spool.

These and other objects and benefits of the present invention will become apparent from the following detailed description thereof taken in conjunction with the accompanying drawings, wherein like reference numerals designate like elements throughout the several views.

DESCRIPTION OF THE DRAWINGS

Fig. 1 is a perspective view of an embodiment of the volumetric feed control in place on a machine;

Fig. 2 is a view of an embodiment of the sensing system of Fig. 1 , taken along line 2-2 thereof;

Fig. 3 is a view of an alternative embodiment of the sensing system of Fig. 2; Fig. 4 is a view of another alternative embodiment of the sensing system of Fig. 2; Fig. 5 is a view of yet another alternative embodiment of the sensing system of Fig. 2; and Fig. 6 is a partial block diagram of a control system for the volumetric feed control.

DESCRIPTION OF THE PREFERRED EMBODIMENT

Referring now to the drawings, Fig. 1 shows the volumetric feed control 10 in place on a three-dimensional modeling machine 12, the machine 12 having an element advancement mechanism such as feed rollers 14, a dispensing head 16 with an application tip 17, and a frame 18. A build material element such as filament 20 is fed to machine 12 from a build element source, not shown, which may be a spool or roll of material, or another such storage device. Build material is often formed as a filament such as filament 20, but may take other configurations. Such other configurations include those of different cross sections. Further, the build material element may be supplied in discrete quantities, and need not be wound in a roll or on a spool.

Volumetric feed control 10 may include dimension control ring 22 and DC servo motor 24 with encoder feedback. Dimension control ring 22 contains sensing means 23 to determine the effective cross section of a build material element such as filament 20 being fed through the dimension control ring 22.

Cross sectional area is often used to estimate volume. Also used are diameter measurements. These methods of determining volume only work when no anomalies are present. Such methods of measurement, especially using conventional instruments and techniques, fail to take into account that if only a diameter measurement is made, a volume computation is meaningless if the element being measured is hollow. The effective cross section is a measurement that takes into account such factors as the cross sectional area of the element, known quantities such as length, width, and diameter, and other information, such as whether the element is hollow, oblong, or the like. The use of effective cross section reduces the possibility for miscalculation due to various physical factors of the element. A more accurate volume calculation may then be made. The sensing means 23 is operatively connected to a central processing unit 74, and continuously gathers data and feeds information signals measuring the build material element being fed therethrough to the central processing unit 74. The central processing unit is also operatively connected to a motor 24, which may be a DC servo motor or a stepper motor. Motor 24 is in turn operatively connected to feed rollers 14 or an other element advancement mechanism which may be used on the modeling machine. The central processing unit 74 controls the speed at which the motor 24 turns, and therefore the speed at which the advancement mechanism rotates to pull the build material element through the dimension control ring 22. Central processing unit 74 adjusts the speed of motor 24 in order to provide a constant flow rate of build material element to dispensing head 16 and application tip 17.

Sensing means 23 need not be housed in a dimension control ring 22. It may be mounted at any place in which it will be positioned to make measurements on element 20.

Sensing means 23 is preferably positioned between any driving contact means such as feed rollers 14 and the application tip 17. This is because the driving contact means may affect the effective cross section of the element due to forces imparted by the feed rollers 14 or other mechanical driving means. Sensing means 23 may even be incoφorated into the element advancement mechanism, such as by mounting the feed rollers to communicate with a linear variable differential transformer as described below.

Referring now also to Figs. 2-4, various embodiments of the sensing means 23 may be seen. In Fig. 2, filament 20 is shown substantially centered in dimension control ring 22. As has been mentioned, dimension control ring 22 is not required. The sensing means 23 may

be mounted elsewhere. Although in this configuration it is preferred that filament 20 be centered in dimension control ring 22, it is not necessary. In Fig. 2, a tungsten halogen lamp 26 is oriented with its lamp filament 28 parallel to and a fixed predetermined distance 30 from the axis 32 of filament 20. The emitted light 34 from the lamp 26 creates a shadow of the filament 20 on a linear CCD array 36. The shadow width is designated as letter A (Fig. 2). Shadows of different widths A will cause a variation in the number of array pixels 38 that are illuminated by the light 34 from lamp 26. The width 21 of filament 20 may then be determined using standard clocking and preamplification techniques known in the art and not further described herein. These measurements are used to generate an effective cross section of filament 20. Referring to Fig. 3, a sensing means 23 using a three-coil linear variable differential transformer (LVDT) 40 and a pair of dimension rollers 42 is shown. Dimension rollers 42 are mounted to shafts 44. Each shaft 44 is flexibly attached to rigid mount 46 by a suitable mounting apparatus, such as pins 48. Rigid mount 46 is attached to the interior surface 72 of the dimension control ring 22, but may also be attached at any fixed point on the frame of the modeling machine. The dimension rollers 42 are biased toward each other by spring 50 attached between shafts 44. Each shaft 44 also carries an attachment for LVDT 40 which is attached therebetween. The terminals of LVDT 40 are operatively connected to rollers 42, one terminal to each roller. As has been described above, tf oilers 42 may bc uscd with multiple functions, such as to also serve as feed rollers for eicment advancement, or encoding the element velocity. Filament 20 passes between dimension rollers 42. As the filament diameter changes, the dimension rollers 42 respond by moving, causing motion of the shafts 44. The movements of the dimension rollers 42 cause the core 52 of LVDT 40 to change position within LVDT

40. Sense circuits known in the art may be employed to translate the relative changes in core 52 position and inductance in the LVDT coils to an analog voltage which may be used to compute the effective cross section of the filament 20.

Fig. 4 shows a third embodiment of the sensing means 23 used in the dimension control ring 22 or alternatively mountable directly to the modeling machine. In this embodiment, the filament effective cross section is calculated using information from a capacitive sensor 54.

The sensor 54 is comprised of a pair of conducting plates 56 sandwiching a thin insulator 58 there between. A typical insulator 58 is mylar, approximately 0.002 inches in thickness. Insulator 58 lies in a plane 60 through which a hole 62 is drilled, the hole 62 being coaxial with the axis 32 of build filament 20 and coplanar with plane 60, and exposing the conductive plates 56. The diameter 64 of hole 62 is slightly larger than the diameter of the largest filament 20 expected to be used with the modeling machine. A capacitance meter 66 is attached to the conductive plates 56. Changing diameter of the filament 20 will cause a change in capacitance that may be translated in known fashion to the cross sectional area of the filament. This method of determining the effective cross section of the filament 20 is independent of its cross sectional shape, since the capacitive sensor 54 essentially measures the percentage of the hole 62 that is filled with filament 20 as opposed to air. The capacitance measurement scheme may also be used to detect the presence of

absorbed water in the filament 20, since the filament 20 material and water have different capacitive properties. Yet another alternative embodiment (Fig. 5) of the sensing means 23 comprises at least one set of emitter/ detector pairs 68, 70 spaced 180 degrees apart around the axis of the filament 20. The emitter/detector pair 68, 70 is situated so that the filament 20 will pass through the beam from emitter 68 as the filament 20 is advanced by feed rollers 14. The emitter 68 sends out a light signal or beam that may be partially occluded by filament 20 and which is detected by the opposite detector 70. If the filament 20 changes in size, the intensity of the beam detected by the detector 70 will change, allowing a measurement which may be used to calculate the filament effective cross section. Preferably, a plurality of emitter/detector pairs are spaced at equal intervals around the axis of filament 20 in order to more accurately detect and measure the effective cross section of the filament 20 as it is advanced by the advancement mechanism. Other types of sensors are also compatible with the volumetric feed control 10, and could be used in place of the emitter/detector pairs. Such other sensors include but are not limited to opto-electric sensors, miniature CCD cameras, fiber optic sensors, mechanical bladder ring sensors, or pinch roller sensors. All could send information that may be used to determine the effective cross section of the flexible filament to the central processing unit 74. The various sensing means 23 described above each send a signal corresponding to the filament effective cross section to central processing unit 74. The central processing unit 74 computes the effective cross section and the required feed speed of the filament 20 to provide a constant flow rate of build material to the application tip 17 of the dispensing head 16, and adjusts the speed of motor 24 which controls the speed of filament feed by controlling the speed of the advancement mechanism such as feed rollers 14. This adjustment ensures that any changes in the effective cross section of the filament 20 will result in a corresponding change in the speed at which the filament 20 is fed. If further accuracy of effective cross section measurement and testing of properties of the filament 20 is desired, two or more sensing means may be combined or used in sequence to determine multiple properties of the filament 20 and provide a reference for averaging computed values. For example, an LVDT sensing means as described above may be used along with a capacitive sensor as also described above, in order to determine both the outer dimension and the water content of the filament 20. Consequently, the tolerance of a filament with nominal diameter of 0.070 inches can be increased substantially from +/- 0.0015 inches.

Another way to increase the accuracy of the effective cross section is to more closely monitor the effective velocity of the element. Although the rotational speed imparted by the element advancement mechanism provides a good approximation of the element velocity, the approximation may be in error due to a number of factors. The element advancement mechanism itself often requires a groove such as that shown in Fig. 3 on rollers 42 in order to seat the build material element properly. The angular velocity of the rollers and accordingly the linear velocity imparted by the rollers differs at the outer periphery and at the central most portion of the groove. To combat such inaccuracies, the element advancement mechanism may be encoded for element velocity. When such an account is made of effective element velocity, this measure may be combined with effective cross section to more accurately compute the volumetric flow rate. Build material elements may be fabricated in a variety of cross sectional shapes and geometries, including ribbons, tubes, extrusions of triangles, trapezoids, pentagons, and other polyhedra. The measurement techniques described above may be adapted to these alternative geometries. In operation, the volumetric feed control 10 works as follows. The sensing means 23 sends measurement signals to central processing unit 74. Central processing unit 74 uses the signals to continuously compute the effective cross section of the build element. In response to this known quantity, central processing unit 74 is able to adjust the feed speed of the build element by adjusting the speed of motor 24 controlling the feed rollers 14 that advance the build element toward the dispensing head 16 and application tip 17. The adjustment is made so that a constant flow rate of build material is fed to the dispensing head

16. A time lag exists between the computation of build element effective cross section and the arrival of the particular measured portion of the element at the application tip

17. Consequently, a change in effective cross section of the build element at the sensing means 23

does not immediately translate to a change in volume of material at the application tip 17. The

change in build element effective cross section will affect the volume of material at the

application tip at some point in time after it passes the sensing means 23.

To account for this lag, the central

processing unit 74 must know the distance 76

between the application tip 17 and the sensing means 23, as well as the distance 78 between the sensing means 23 and the center of feed rollers 14. The connection of motor 24 to the central processing unit allows it to know or compute at all times both the effective cross section of the build element and the speed of the feed rollers 14. The central processing unit

74 can therefore correct for the lag response in the system and provide a constant volume of material at the application tip 17 regardless of variances

in the effective cross section of the build element.

The detailed description outlined above is considered to be illustrative only of the principles of the invention. Numerous changes and modifications will occur to those skilled in the art, and there is no intention to restrict the scope of the invention to the detailed description. The preferred embodiment of the invention having been described in detail, the

scope of the invention should be defined by the following claims.

Automatisches Kalibrierungssystem

Veröffentlichungsnummer	US20070228592 A1
Publikationstyp	Anmeldung
Anmeldenummer	US 11/397,012
Veröffentlichungsdatum	4. Okt. 2007
Eingetragen	3. Apr. 2006
Prioritätsdatum	3. Apr. 2006
Auch veröffentlicht unter	CN101460290A, 5 weitere »
Erfinder	James Comb, Benjamin Dunn, Hans Erickson,Jason Wanzek
Ursprünglich Bevollmächtigter	Stratasys, Inc.

Referenziert von (8), Klassifizierungen (6), Legal Events (2)

Externe Links: USPTO, USPTO-Zuordnung, Espacenet

Kommentar:

Die zweiteilige automatische Kalibrierung, besteht aus einem Normobjekt, das in eine definierte Position der Druckfläche gedruckt wird und einer Prüfspitze, die das Objekt danach in den drei Raumkoordinaten vermisst und die Abweichungen an die Software als Versatz zurückmeldet. Eine bestechend einfache Lösung. Die hier gezeigt mechanische Messspitze kann natürlich auch durch andere Messverfahren (optisch usw.) ersetzt werden.

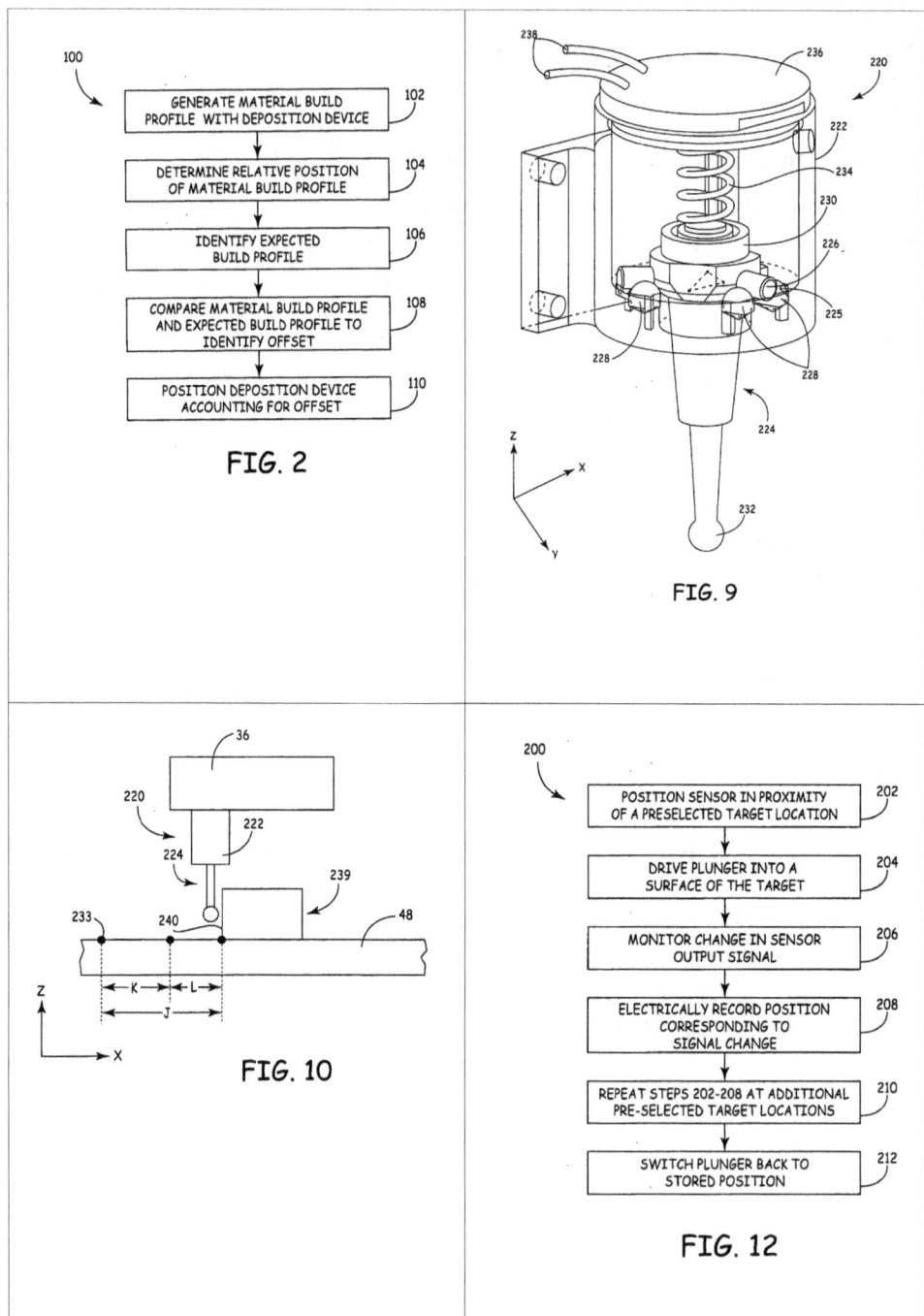

FIG. 2

FIG. 9

FIG. 10

FIG. 12

Spritzguss mit FDM Drucker

Titel"Rapid prototype injection molding"

Veröffentlichungsnummer	US20050173839 A1
Publikationstyp	Anmeldung
Anmeldenummer	US 10/511,787
PCT-Nummer	PCT/US2003/011854
Veröffentlichungsdatum	11. Aug. 2005
Eingetragen	17. Apr. 2003
Prioritätsdatum	17. Apr. 2002
Auch veröffentlicht unter	CN1652915A, 6 weitere »
Erfinder	Steven Crump, William Priedeman Jr,Jeffery Hanson
Ursprünglich Bevollmächtigter	Stratasys, Inc.

Referenziert von (2), Klassifizierungen (17), Legal Events (2)

Externe Links: USPTO, USPTO-Zuordnung, Espacenet

Kommentar:

Das Verfahren beschreibt die Nutzung eines FDM Druckers um in zwei Schritten ein Spritzgussmodell zu erhalten. Beschrieben werden zunächst Patente, die dies für Stereolithographiemaschinen ermöglichen. Hier wird zunächst die Form in Sl erzeugt und dann mit einer thermisch leitenden Schicht versehen.

Bei diesem Patent hier wird zunächst wir eine Form per FDM hergestellt. Diese Verfügt über eine Eingussöffnung, Entlüftungsbohrungen sowie 4 Zentrierpassungen um die beiden Formhälften passgenau zusammen zu setzen. Das Patent nennt PEEK und vergleichbare Materialien für die Form. Weiterhin gibt es Angaben zu dem Prozessparametern. So soll ein FDM Drucker dazu benutzt werden, um in die Öffnung der Form **langsam und mit geringem Druck** Material einzuspeisen. Dies sei der wesentliche Unterschied zu einem echten Spritzgussverfahren.

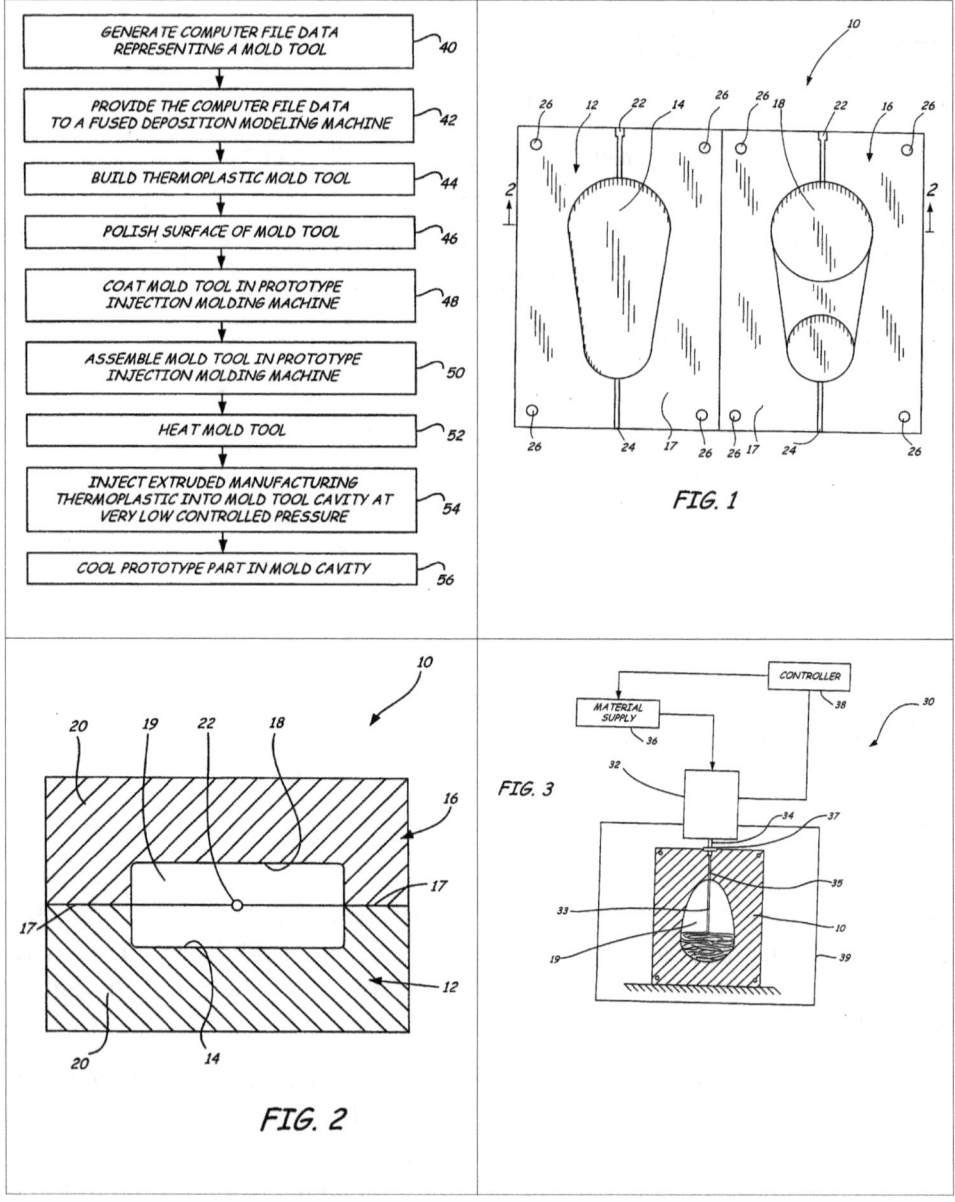

GENERATE COMPUTER FILE DATA REPRESENTING A MOLD TOOL — 40

PROVIDE THE COMPUTER FILE DATA TO A FUSED DEPOSITION MODELING MACHINE — 42

BUILD THERMOPLASTIC MOLD TOOL — 44

POLISH SURFACE OF MOLD TOOL — 46

COAT MOLD TOOL IN PROTOTYPE INJECTION MOLDING MACHINE — 48

ASSEMBLE MOLD TOOL IN PROTOTYPE INJECTION MOLDING MACHINE — 50

HEAT MOLD TOOL — 52

INJECT EXTRUDED MANUFACTURING THERMOPLASTIC INTO MOLD TOOL CAVITY AT VERY LOW CONTROLLED PRESSURE — 54

COOL PROTOTYPE PART IN MOLD CAVITY — 56

FIG. 1

FIG. 2

FIG. 3

CONTROLLER

MATERIAL SUPPLY

The exemplary mold tool 10 is formed from a non-conductive thermoplastic material that will sustain the temperature and pressure of the injection molding process, so as to produce at least one prototype plastic injection molded part. An exemplary thermoplastic comprises at least 50 weight percent of a thermoplastic selected from the group consisting of polycarbonate, polystyrene, acrylics, amorphous polyamides, polyesters, polyphenylsulfone, polysulfone, polyphenylene ether, nylon, PEEK, PEAK, poly(2-ethyl-2-oxazoline), and blends thereof. The thermoplastic resin may contain various fillers, additives and the like, as will be understood by those skilled in the art. A particularly preferred thermoplastic for use in creating a mold tool by fused deposition modeling is a polyphenylsulfone-based resin.

FIG. 3 shows an exemplary rapid prototype injection molding apparatus 30 in accordance with the present invention, in the process of making a prototype part. The apparatus 30 comprises an extrusion head 32 having a dispensing tip 34, a material supply 36, a controller 38, a modeling envelope 39, and a hollow sprue 35. The mold tool 10 is assembled and positioned in the apparatus 30. Prior to assembly of the mold tool 10, the mold surfaces 14 and 18 are created with a release agent that facilitates removal of a completed part from the mold tool 10. Suitable release agents include dry film lubricants, and others that will be recognized by those skilled in the art. The mold tool 10 in mounted in the modeling envelope 39. Sprue 35 is placed in the sprue channels 22, such the a dispensing end of the sprue 35 is directed into the mold cavity 19. The sprue 35 has an entry 37 at a top end thereof, designed to mate with the downward-facing extrusion head tip 34. The sprue entry 37 receives and attached to the extrusion head tip 34, thereby providing a flow path from the extrusion head 32 into the mold cavity 19. Preferably, an insulator is provided for the extrusion head tip 34, so that the tip 34 will not cause melting of the prototype part as it is being formed. Also, a pressure transducer (not shown) is placed in the sprue to monitor pressure in the mold cavity so that a predetermined pressure may be maintained.

The apparatus 30 may be a fused deposition modeling machine. It should be understood, however, that unlike fused deposition modeling, the process of the present invention involves no translational movement of the extrusion head. The extrusion head 32 may be of any type which receives a thermoplastic material and dispenses the material in a molten state through a dispensing tip at low flow rates and low pressure. Suitable extrusion heads have been developed for fused deposition modeling, and include a liquifier pump, a piston pump and a screw pump. Each of these extrusion heads developed for three-dimensional modeling receives a feedstock of thermoplastic in solid form, and heats the thermoplastic material to a desired temperature for extrusion.

In the exemplary embodiment, the extrusion head 32 receives a supply of production thermoplastic material for creating the molded prototype part from the material supply 36, at a rate controlled by the controller 38. Where the extrusion head 32 of the exemplary embodiment is a liquifier pump, the material supply 36 comprises spooled flexible filament and the extrusion head 32 carries a set of feed rollers for advancing the filament into the extrusion head at the controlled rate. Liquifier pumps are disclosed, for example, in U.S. Pat. No. 6,004,124. Where the extrusion head 32 of the exemplary embodiment is a piston pump, the material supply 36 comprises cylindrical feed rods of thermoplastic material fed in a batch process. A piston pump extrusion head is disclosed in U.S. Pat. No. 6,067,480. Where the extrusion head 32 of the exemplary embodiment is a screw pump, the material supply 36 comprises pellets of thermoplastic material. A screw pump extrusion head is disclosed, for example, in U.S. Pat. No. 5,312,224. Two-stage extrusion heads are also known in the art, and can also be used in practice of the present invention.

92 Ausgewählte Patente

A extrusion pump is disclosed in U.S. Pat. No. 5,764,521, wherein the feedstock received from material supply 36 is pressurized in a two-stage process which may take various forms.

The extrusion head 32 may include an ultrasonic vibrator for creating a thixotropic flow at the dispensing tip exit, such as is disclosed in U.S. Pat. No. 5,121,329. The ultrasonic energy would reduce the injection pressure while increasing the flow rate of the production thermoplastic.

For the production of a prototype part, production thermoplastic is provided from the material supply 36 to the extrusion head 32, which heats the production thermoplastic to an extrusion temperature and dispenses molten extruded material 33 through sprue 35 and into the mold cavity 19. Production thermoplastics that may be used in the present invention include, without limitation, ABS, polycarbonate, polystyrene, acrylics, amorphous polyamides, polyesters, polyphenylsulfone, polyphenylene ether, nylon, PEEK, PEAK, and blends thereof. The production thermoplastic may, of course, include various fillers, additives and the like. Shrink characteristics of the mold tool plastic are matched to the shrink characteristics of the production thermoplastic, which can be achieved by using amorphous thermoplastics. Also, the production thermoplastic must have a heat deflection temperature lower than a heat deflection temperature of the plastic which forms the mold tool, so that the mold tool will maintain its shape.

FIG. 4 shows a flow diagram which summarizes an exemplary method of producing a prototype injection molded part in accordance with the present invention, using a mold tool made by fused deposition modeling. A CAD tool is used to generate computer file data representing a mold tool, in a step 40. The data is provided to a fused deposition modeling machine, in a step 42. The mold tool is built in the fused deposition modeling machine, in layers defined by the computer file data, in a step 44. In a step 46, the mold surfaces and/or mating surfaces of the mold tool are smoothed to remove ridges unintentionally created in the formation of the mold tool.

In a preferred embodiment, the smoothing is done in a vapor smoothing process, which is the subject of International Application No. PCT/US03/_____ entitled "Smoothing Method For Layered Deposition Modeling," filed Apr. 4, 2003, assigned to the same assignee as the present application, and hereby incorporated by reference as if set forth fully herein. As is disclosed in said co-pending application, the surfaces of the mold tool can be smoothed by placing the mold tool in a vaporizer and exposing it to vapors of a solvent until a desired surface finish is obtained. The solvent is selected to be compatible with the material which forms the mold tool. Suitable solvents will react with the material so as to soften and flow the material at the object surfaces. A preferred solvent for use with a wide range of amorphous thermoplastics is methylene chloride. Other suitable solvents will be recognized by those skilled in the art, for instance, an n-Propyl bromide solution (e.g., Abzol®)), perchloroethylene, trichloroethylene, and a hydrofluorocarbon fluid sold under the name Vertrel®. Vapor smoothing will also serve to seal the surfaces of the mold tool. As is taught in said co-pending application, certain mold features may be identified for solvent masking or for pre-distortion prior to the vapor smoothing step, and the computer file data representing the mold tool may include data identifying said features. Alternatively, smoothing can be done by applying a liquid solvent. Other alternative smoothing techniques include sanding, grinding, and thermal ironing.

The mold surfaces of the mold tool are then coated with a release agent, in a step 48 (a soluble mold tool may not need a release agent). Suitable release agents include dry film lubricants, and others that will be recognized by those skilled in the art. If needed, sprue and vent channels and alignment holes are machined into the mold tool prior to step 48. The mold tool is assembled in an prototype injection molding apparatus according to the present invention, without the addition of any conductive fill material or layers, in a step 50. The fused deposition modeling machine used to build the mold tool may be used also as the prototype injection molding apparatus.

In step 50, the sprue is positioned in the mold tool and attached to a dispensing tip of an extrusion head. The mold may be clamped to a fixture to hold it in place, using a clamp or other means. A clamping force of less than or equal to about 10 tons will ensure office compatibility of the injection process. The mold tool is then heated to a predetermined temperature, in a step 52.

Injection molding is then performed, in a step 54. Production thermoplastic is injected from an extrusion head of the molding apparatus into the mold tool, at low speed and low pressure. If an ultrasonic vibrator is used in the extrusion head, a thixotropic flow of thermoplastic will be injected. During the injection, pressure is preferably monitored, such as by a pressure transducer placed in the mold cavity or sprue. A system controller can adjust the extrusion flow rate based upon pressure reading from the

transducer, to maintain the pressure within a target range. The target pressure will typically be less than 5000 psi and may be set at less than 500 psi, to as low as less than 20 psi (near-zero pressure).

Filling the mold cavity will typically take about 1-2 hours. During this time, the temperature of the mold and the production thermoplastic remain approximately constant, as the plastic mold is a thermal insulator having a high thermal resistance. In the exemplary embodiment, the mold tool is heated prior to the injection step 54, to provide isothermic conditions. Various techniques may be used to heat the mold, as will be recognized by those skilled in the art. A temperature sensor may be placed in the mold cavity, and temperature of the thermoplastic flow, the mold tool, and/or the build environment can then be adjusted as needed to maintain isothermic conditions. Maintaining the temperature of the mold tool at approximately the extrusion temperature of the production thermoplastic prevents premature solidification of the production thermoplastic, so that the mold cavity can fill completely before the prototype part hardens. It should be understood, however, that some materials and process parameters may provide isothermic conditions without the need for pre-heating the mold, in which case step 52 may be omitted. Also, it may be desirable to heat the mold cavity and inner mold surface, rather than heating the entire mold.

During (and prior to) the injection step 55, a vacuum may be drawn upon the mold tool. This may be done, for example, by placing the mold tool in a vacuum chamber. The vacuum will remove gases from the mold tool and the sprue, through the porosity of the mold or the vent. The vacuum will assist in pulling the injected material into the mold cavity. The vacuum will also facilitate filling the mold cavity more fully and consistently than would be achieved under normal atmospheric conditions, resulting in a void-free part.

When the mold cavity is filled, injection of material into the mold cavity is terminated and the prototype part is allowed to cool inside the mold tool, in a step 56. Cooling may take from 1 to 2 hours. Pressure may be placed on the mold tool during cooling, to compensate for shrinkage of the mold tool and the prototype part. If desired, active cooling may be used to speed the cooling process. Also, monitoring of pressure in the mold cavity may be continued during cooling. Cooling in the mold tool may be allowed to continue until the prototype part reaches room temperature, or the part may be removed when it is substantially cool.

Using the method of the present invention, a prototype plastic injection molded part can be produced within a 24-hour time period. Up to 50 prototype injection molded parts could be produced within 48 hours.

It should be understood that a mold tool for use in carrying out the present invention need not be limited to a thermoplastic mold build by a deposition modeling process. Rather, any non-conductive plastic mold tool compatible with the injection process may be utilized, including, for example, mold tools formed by stereolithography, thermoset mold tools, and mold tools formed by a machine-removable process (e.g., CAM/CNC). A plastic material forming the mold tool may include various fillers, additives and the like.

Although the present invention has been described with reference to preferred embodiments, the invention is defined by the claims. Workers skilled in the art will recognize that changes may be made in form and detail without departing from the spirit and scope of the invention.

For example, in one alternate embodiment, the mold tool is transparent and a photopolymer is used to form the prototype part. The photopolymer is injected into the mold cavity and cured by exposure to light. In another embodiment, thermoset reaction injection molding techniques are employed. Two or more reactant materials are mixed together to form a thermoset resin, which is injected into the mold tool. The resin is cured to form the prototype part by applying heat to the mold tool.

Schmelzkompensation eines Filaments

Veröffentlichungsnummer	US6547995 B1
Publikationstyp	Erteilung
Anmeldenummer	US 09/960,133
Veröffentlichungsdatum	15. Apr. 2003
Eingetragen	21. Sept. 2001
Prioritätsdatum	21. Sept. 2001
Gebührenstatus	Bezahlt
Auch veröffentlicht unter	CN1291829C, CN1555306A, EP1427578 A1,EP1427578A4, EP1427578B1,US200 30064124, WO2003026872A1,
Erfinder	James W. Comb
Ursprünglich Bevollmächtigter	Stratasys, Inc.

Patentzitate (7), Referenziert von (30), Klassifizierungen (12),Legal Events (10)

Externe Links: USPTO, USPTO-Zuordnung, Espacenet

Kommentar:

Wird das Filament im Extruder bis zum Schmelzpunkt erhitzt, dehnt es sich aus. Wird es dann extrudiert und kühlt ab schrumpft es wieder. Das zu dosierende Volumen ist daher ungleich des Volumens, dass sich durch die Länge des geförderten Filaments errechnen lässt. Die Erfindung beschreibt daher ein Formelwerk, dass diese Abweichung einbezieht und die dosierte Menge vorab um diese Volumenänderung korrigiert. Die Korrekturalgorithmen müssen dann in die Betriebssoftware des Systems integriert werden.

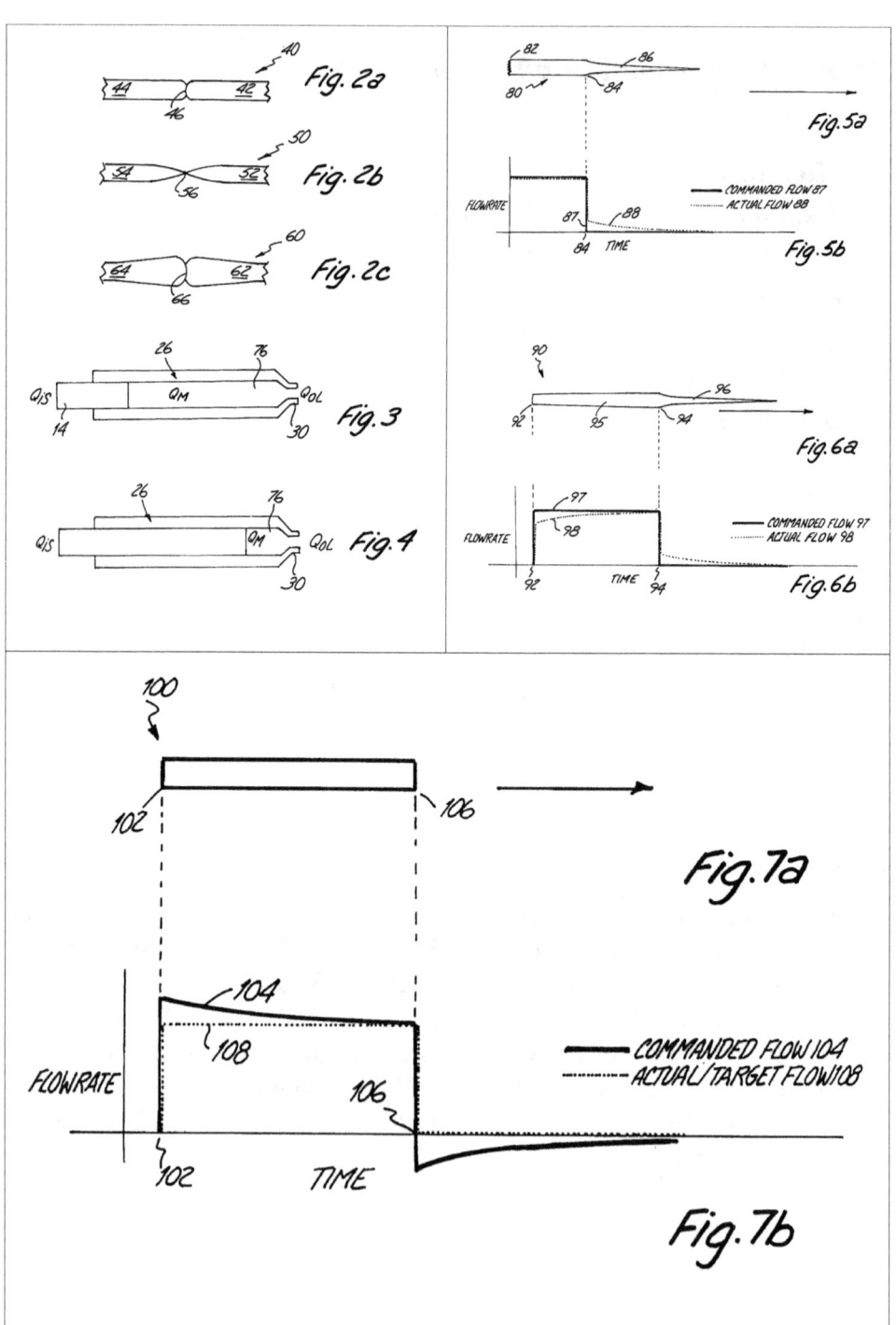

Fig. 2a

Fig. 2b

Fig. 2c

Fig. 3

Fig. 4

Fig. 5a

Fig. 5b

COMMANDED FLOW 87
ACTUAL FLOW 88

Fig. 6a

Fig. 6b

COMMANDED FLOW 97
ACTUAL FLOW 98

Fig. 7a

Fig. 7b

FLOWRATE

TIME

COMMANDED FLOW 104
ACTUAL/TARGET FLOW 108

BACKGROUND OF THE INVENTION

This invention relates to the fabrication of three-dimensional objects using extrusion-based layered manufacturing techniques. More particularly, the invention relates to supplying solid modeling material to a liquifier carried by an extrusion head, and extruding the material in a flowable state in a predetermined pattern in three dimensions with respect to a base.(..)

One type of rapid prototyping system of the prior art drives the motion of the extrusion head at a constant velocity along a tool path comprising a poly-line. A poly-line is a continuous curve of straight-line segments defined by a list of X-Y coordinate pairs at each vertex. The head velocity is preselected so as to accomplish the general goal of moving the extrusion head quickly along the poly-line while minimizing the displacement from the tool path. As a result, the head velocity must be set to be slow enough that the deviation will not exceed the maximum allowable following error for the largest deflection along that poly-line. Using a constant head velocity along a tool path, bead width remains fairly constant but errors arise at start points and end points of the tool path, for instance, at the location of a "seam" (i.e., the start and end point of a closed-loop tool path).

Another type of prototyping system of the prior art varies the extrusion head speed to increase the throughput of the modeling machine. The extrusion head speeds up along straight-aways in the tool path, and slows down where there are deflection angles or vertices. U.S. Pat. No. 6,054,077 describes one such technique for varying the extrusion head speed, using X-Y trajectory profiling that follows the exponential step response of the liquifier pump. The velocity profile of the extrusion head looks like a "shark tooth", while the pump profile follows a step function.

It has been observed that the variable velocity systems of the prior art introduce greater bead width error, and also have seam errors. It would be desirable to reduce errors in bead width and seam quality so as to achieve a desired extrusion profile, while allowing the higher throughput of a variable rate system.

BRIEF SUMMARY OF THE INVENTION

The present invention is a liquifier pump control method and apparatus which reduces bead width errors and seams errors observed in the prior art by accounting for thermal expansion of the modeling material in the liquifier. The melting of modeling material is accompanied by its expansion. The present invention recognizes that the melt expansion produces unanticipated extruded flow rates from the liquifier during transient conditions. The present invention predicts a melt flow component of the extruded flow rate produced by the thermal expansion of the modeling material, and compensates for the predicted melt flow in a commanded flow rate.

BRIEF DESCRIPTION OF THE DRAWINGS

FIG. 3 is a graphical representation of a liquifier operating at a minimum flow rate.

FIG. 4 is a graphical representation of a liquifier operating at a maximum flow rate.

FIG. 5a is a view of an extrusion profile extruded by a prior art liquifier pump operating at a steady state and then turned off.

FIG. 5b is a graphical representation of the amount of flow produced by the liquifier pump of FIG. 5a.

(..)

The present invention recognizes that melt expansion of the modeling material is a significant cause of errors in the desired extrusion profile, such as the seam errors illustrated in FIGS. 2b and 2 c. Utilizing the present invention, the melt flow component of the extruded flow rate is predicted and is compensated for by adjusting the input rate of solid material, resulting in significantly reduced errors in bead width and seams.

The melt flow compensation of the present invention takes into account the flow history of the liquifier to command a flow rate that will account for melt flow. FIG. 3 is a graphical representation of a cross-section of the liquifier 26 operating at a minimum flow rate. FIG. 4 is a graphical representation of a cross-section of liquifier 26 operating at a maximum flow rate. The filament 14 is fed into the liquifier 26 at an input (or commanded) flow rate QiS, heated in the liquifier 26 to a liquid 76 at a melting rate QM, and extruded out of the tip 30 of liquifier 26 at an output flow rate QoL. As illustrated, at higher flow rates, more of the liquifier 26 contains solid modeling material in the form of filament 14, as compared to melted modeling material (liquid 76). This is due to limited melt capacity of the liquifier. If the input flow rate, QiS, were to go from a higher to a lower rate, the amount of liquid 76 in the liquifier 26 will increase and the output flow rate, QoL, will include a melt flow component, QMFL, that is taken into account by the present invention by way of a downward adjustment of the commanded input flow rate, QiS.

In order to account for melt flow, the melt flow characteristic of a given operating system may be modeled by an equation. The melt rate of a solid rod of material in a cylindrical liquifier has been observed to be approximately exponential. For a step increase in solid material input rate, the rate of melting increases exponentially to an asymptotic value equal to the input rate of solid material. When the liquifier pump is turned on, the melt flow rate of material from the liquifier increases approximately exponentially. Conversely, when the liquifier pump is turned off, the melt flow rate exponentially decreases to zero. Accordingly, melt flow can be predicted by an exponential equation dependent upon a melt flow time constant of the liquifier.

(..)

What is claimed is:

1. In an extrusion apparatus having a liquifier which receives a solid element of a material that exhibits thermal expansion, heats the material, and deposits a flow of the material through a dispensing tip thereof along a predetermined tool path at an output rate, said apparatus using a material advance mechanism to supply the solid element of material to the liquifier at an input rate which controls the output rate, a method for matching the output rate to a predetermined target output rate which is selected to achieve a desired extrusion profile of the material deposited along the tool path, comprising the steps of:

predicting a melt flow component (QMF) of the output rate for a time step corresponding to a segment of the tool path, the melt flow being a rate of flow attributed to thermal expansion of the material heated in the liquifier; and

commanding the input rate (QiS) for that

time step so as to compensate for the predicted melt flow.

2. The method of claim 1, and further comprising the step of:

repeating the steps of predicting and commanding for subsequent time steps.

3. The method of claim 1, wherein the melt flow is predicted as a function of a melt flow time constant of the liquifier (τMF) and a percent thermal expansion of the material (%MF).

4. The method of claim 3, wherein predicting the melt flow component (QMF) of the output rate at the time step comprises adding the melt flow component from a previous time step (QMF t−1) to a predicted change in the melt flow component from the previous time step (ΔQMF), and wherein the input rate (QiS) for the time step is commanded according to the equation QiS=(1+%MF) (QTarget−QMF) where Qtarget is the predetermined target output rate.

5. The method of claim 4, wherein the predicted change in the melt flow component is given by the difference equation

ΔQ MFS = % MF $\star$ Q iS t - 1 1 + % MF - Q MFS t - 1 $\star \Delta$ t τ MF ,

$$\Delta Q_{MFS} = \frac{\%_0{}_{MF} * Q_{iS_{t-1}}}{1 + \%_0{}_{MF}} - Q_{MFS_{t-1}} * \frac{\Delta t}{\tau_{MF}}.$$

where QiS t−1 is the input rate from the previous step, and Δt is the duration of a time step.

6. The method of claim 1, wherein the melt flow over time is predicted using an exponential model.

7. The method of claim 6, wherein the exponential model is a function of a melt flow time constant of the liquifier (τMF) and a percent thermal expansion of the

material (%MF).

8. The method of claim 7, wherein the exponential model of melt flow over time is given by the equation

Q MFS = % MF $\star$ Q iS 1 + % MF $\star$ (1 - イ - t τ MF),

$$Q_{MFS} = \frac{\%_0{}_{MF} * Q_{iS}}{1 + \%_0{}_{MF}} * \left(1 - e^{\frac{-t}{\tau_{MF}}}\right).$$

for a step change in QiS from zero.

9. An extrusion apparatus comprising:

a liquifier which receives a solid element of a material that exhibits thermal expansion, heats the material, and deposits a flow of the material through a dispensing tip thereof along a predetermined tool path at an output rate;

a material advance mechanism which supplies the solid element of material to the liquifier at an input rate (QiS) that controls the output rate;

a control for providing control signals to the material advance mechanism, the control signals commanding operation of the material advance mechanism so that the input rate compensates for a predicted melt flow component (QMF) of the output rate.

10. The extrusion apparatus of claim 9, wherein the control contains an algorithm for predicting the melt flow component as a function of a melt flow time constant of the liquifier (τMF) and a percent thermal expansion of the material (%MF).

11. The extrusion apparatus of claim 10, wherein the algorithm comprises calculating a predicted change in the melt flow component (QMF) from a previous time step to a next time step according to the difference equation

ΔQ MFS = % MF ⋆Q iS t - 1 1 + % MF - Q MFS t - 1 ⋆Δ t τ MF ,

$$\Delta Q_{MFS} = \frac{\%o_{MF} * Q_{iS_{t-1}}}{1 + \%o_{MF}} - Q_{MFS_{t-1}} * \frac{\Delta t}{\tau_{MF}}.$$

where QiS t−1 is the input rate at the previous time step, QMFS t−1 is the melt flow component of the output rate of solid material at the previous time step, and Δt is the duration of a time step.

12. The extrusion apparatus of claim 9, where melt flow over time is predicted using an exponential model.

13. The extrusion apparatus of claim 12, wherein the exponential model of melt flow over time is given by the equation

Q MFS = % MF ⋆Q iS 1 + % MF ⋆(1 - t τ MF) ·

$$Q_{MFS} = \frac{\%o_{MF} * Q_{iS}}{1 + \%o_{MF}} * \left(1 - e^{\frac{-t}{\tau_{MF}}} \right).$$

Dispenser

3D Druck unter Wasser

Titel: "Vorrichtung und Verfahren zum Herstellen von dreidimensionalen Objekten"

Veröffentlichungsnummer	DE10018987 A1
Publikationstyp	Anmeldung
Anmeldenummer	DE2000118987
Veröffentlichungsdatum	31. Okt. 2001
Eingetragen	17. Apr. 2000
Prioritätsdatum	17. Apr. 2000
Auch veröffentlicht unter	CN1450953A, 6 weitere »
Erfinder	Hendrik John, Ruediger Landers, Rolf Muelhaupt
Antragsteller	Envision Technologies Gmbh

Patentzitate (1), Referenziert von (8), Klassifizierungen (12),Legal Events (2)

Externe Links: DPMA, Espacenet

Kommentar:

Der zunächst etwas sperrige Patenttext, erweist sich bei genauerem Hinsehen als interessantes Verfahren um mittels Dispenser ein schnell aushärtendes Modell zu erzeugen. Das Prinzip liegt einem chemischen Mechanismus zugrunde, nachdem manche Silikone (und eventuell auch andere Stoffe) bei Kontakt mit Wasser sofort aushärten. Das macht sich das Verfahren zunutze, indem die Nadel des Dispensers unter Wasser arbeitet. Sobald also das Druckmaterial den Kolben verlässt wird es in kurzer Zeit so fest, dass das fertige Modell Gebrauchswert hat. Die Erfinder fügen zudem eine Möglichkeit hinzu, die Eigenschaften des erzeugten Materials durch Dotierung mit einem weiteren Stoff noch im Druckkolben zu beeinflussen.

Auf Stützstrukturen kann angeblich fast immer verzichtet werden, da das Material so schnell aushärtet, dass es Überhänge sofort tragen kann.

Als Material wird nass aushärtendes Silikon oder Silikonkautschuk erwähnt.

Es ist davon die Rede, lebende menschliche Zellen in das Objekt einzubauen.

Das Arbeiten unter Wasser soll zudem durch seine Auftriebskraft, das Zerfließen der gesetzten Punkte verhindern und somit eine bessere Druckqualität liefern.

Dem Autor stellt sich jedoch die Frage, ob es gewährleistet ist, dass die Nadel des Dispensers sich nicht sofort durch diesen Härteprozess zusetzt und unbrauchbar wird. Am interessantesten erscheinen die Ansprüche, da hier die chemische Zusammensetzung der Materialien und Härter angegeben werden.

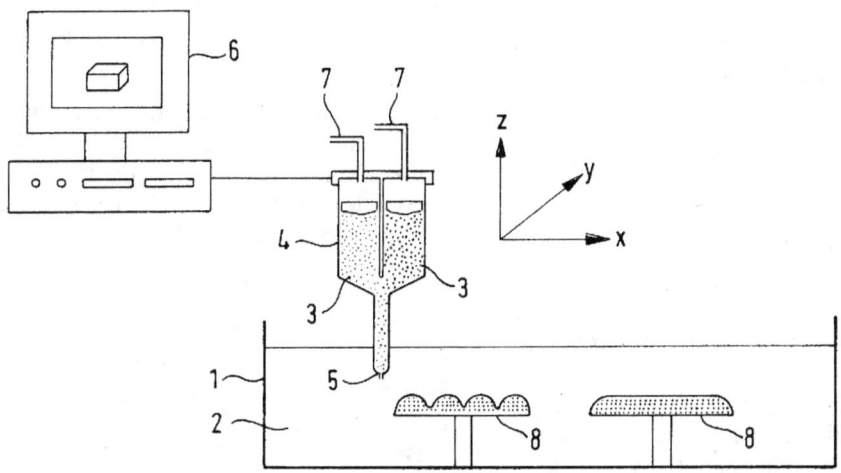

BESCHREIBUNG

(..)

Bei dem erfindungsgemäßen Verfahren wird eine Ausgangsöffnung eines dreidimensional bewegbaren Dispensers in einem ersten Material (2) - dem Plotmedium - positioniert und ein aus einer oder mehreren Komponenten bestehendes zweites Material (3), das in Kontakt mit dem ersten Material (2) zur Ausbildung fester Strukturen führt, wird in das erste Material (2) zur Ausbildung

dreidimensionaler Objekte abgegeben.

Im folgenden wird das erste Material 2 als Medium bzw. Plot medium 2 und das zweite Material 3 als Material 3 bezeichnet, um eine bessere Unterscheidung zwischen erstem (2) und zwei tem (3) Material vornehmen zu können.

Die Wirkung des Mediums (2) besteht dabei zum einen in einer Auftriebskompensation als auch in einer Dämpfung der Bewegung des dosierten, noch flüssigen Materials (3).

Die beiden Effekte zeigen sich deutlich in den Fig. 2 und 3. Bei Fig. 2 führt die mangelnde Auftriebskompensation zu einem Verfließen der dreidimensionalen Gitterstruktur des Datensatzes. Dagegen ist bei Fig. 3 die Gitterstruktur gut ausgebildet und die Hohlraumstruktur zwischen den Schichten bleibt voll erhalten.

Diese technische Änderung, "Plotten" (Dispergieren) eines Materials (3) in ein Medium (2) mit entsprechenden rheologischen, im folgenden näher beschriebenen Eigenschaften, führt zu einer wesentlichen Ausweitung der Breite verwendbarer Materialien.

Zum einen lassen sich das (die) Material(ien) (3) mit geringer Viskosität zu komplizierten dreidimensionalen Objekten aufbauen. Zum anderen kann das Medium (2) in reaktiver Form in den Härtungsprozeß des Materials (3) einbezogen werden. Hierbei können chemische Reaktionen ablaufen, aber auch Fällungs- und Komplexbildungsreaktionen. Die Polarität des Materials 2 variiert abhängig von der Polarität des Materials (3) von hydrophil (z. B. Wasser) bis völlig unpolar (z. B. Silikonöl) um die Haftungseigenschaften der Schichten aneinander zu steuern.

Auf eine Stützkonstruktion kann bei dem hier beschriebenen Verfahren zum Aufbau dreidimensionaler Objekte fast immer verzichtet werden.

Ein sehr wichtiges Detail der Erfindung beruht auf der Temperaturvariabilität des Verfahrens. In Verbindung mit der großen Zahl möglicher Medium (2)/Material (3)-Kombinationen sind Verarbeitungsbedingungen auch bei Raumtemperatur realisierbar.

So können Pharmaka oder lebende, menschliche Zellen in 3D- Strukturen eingebaut werden.

In einer Weiterbildung des Verfahrens wird als Medium (2) Gelatinelösung oder Wasser und als Material (3) Silikon kautschuk verwendet.

In einer Weiterbildung des Verfahrens wird als Medium (2) Wasser und als Material (3) ein naß aushärtbares Silikon mit Acetoxysilangruppen verwendet.

In einer Weiterbildung des Verfahrens wird als Medium (2) Wasser, ein Polyol oder eine Lösung polyfunktioneller Amine und als Material (3) ein Polyurethan (präpolymer) mit Iso cyanatgruppen verwendet.

In einer Weiterbildung des Verfahrens wird als Medium (2) ei ne wässrige Lösung von Calciumionen und Thrombin verwendet, als Material (3) eine wässrige Lösung von Fibrinogen. (..)

ANSPRÜCHE(25)

1. Verfahren zum Herstellen eines dreidimensionalen Objekts mit Bereitstellen eines Mediums (2) in einem Behälter (1), Positionieren einer Ausgangsöffnung (5) eines dreidimen sional bewegbaren Dispensers (4) in dem Medium (2), Abgeben eines aus einer oder mehreren Komponente(n) bestehenden Materials (3) durch den Dispenser (4) in das Medium (2), wobei das Material (3) nach der Abgabe in das Medium (2) aushärtet, oder in Kontakt mit dem Medium (2) zur Ausbildung fester Strukturen führt, und Bewegen des Dispensers (4) an die Stellen, die dem dreidimensionalen Objekt entsprechen, zum Ausbilden einer festen dreidimensionalen Struktur.

2. Verfahren nach Anspruch 1, bei dem das Medium (2) in einer vorbestimmten Füllhöhe in dem Behälter (1)

bereitgestellt wird und die Ausgangsöffnung (5) des Dispensers unterhalb der Füllhöhe des Mediums (2) in dem Behälter (1) positioniert wird.

3. Verfahren nach Anspruch 1 oder 2, bei dem die Dichte des Mediums (2) ungefähr gleich, unwesentlich größer oder kleiner der Dichte des Materials (3) gewählt wird.

4. Verfahren nach einem der Ansprüche 1 bis 3, bei dem aus dem Material (3) Mikrodots auf Lücke, auf Deckung oder spiralförmig gebildet werden, oder ein oder mehrere Mikrostränge gebildet werden, wobei der oder die Mikrostränge individuell oder zusammenhängend, endlos oder portionsweise, spiralförmig gewunden oder linienförmig, mit einem kontinuierlichen oder diskontinuierlichen Materialfluß dosiert werden.

5. Verfahren nach einem der Ansprüche 1 bis 4, bei dem flüssige oder pastöse Komponenten des Materials (3) ver wendet werden, das als Mikrotropfen oder als Mikrostrahl dosiert wird.

6. Verfahren nach einem der Ansprüche 1 bis 5, bei dem das Material (3) als Strang mit einem Kern und einer Schale dosiert wird.

7. Verfahren nach einem der Ansprüche 1 bis 6, bei dem eine Präzipitation des Mediums (2) und/oder des Materials (3) ausgeführt wird, oder

bei dem eine kontrollierte Präzipitation zur Ausbildung von Häuten um Substrukturen der dreidimensionalen Objekte ausgeführt wird, oder

bei dem das Medium (2) ein oder mehrere Fällmittel zum Ausfällen des Materials (3) enthält und das Material (3) ausgefällt wird.

8. Verfahren nach einem der Ansprüche 1 bis 7, bei dem das Material (3) co-reaktive Komponenten enthält, die

miteinander reagieren, und/oder das erste Medium (2) eine co-reaktive Komponente enthält, die mit einer oder mehreren Komponenten des Materials (3) reagiert.

9. Verfahren nach Anspruch 8, bei dem eine Grenzflächenpolymerisation, eine Polykondensation oder eine Polyelektrolytkomplex-Bildung ausgeführt wird.

10. Verfahren nach einem der Ansprüche 1 bis 9, bei dem durch Entfernen des Materials (3) des Kerns eines Kern/Schale-Stranges oder durch Ausführen einer Grenzflächenpolymerisation und Entfernen des Materials (3), das bei der Grenzflächenpolymerisation nicht reagiert hat, Mikrohohlräume oder Mikroröhren gebildet werden.

11. Verfahren nach einem der Ansprüche 1 bis 10, bei dem das erste Medium (2) durch Zudosieren des Materials (3) durch das oder mit dem Material (3) gelöst, gebunden, geschmolzen, gehärtet oder verklebt wird, oder bei dem das Material (3) durch Zudosieren in das Medium (2) durch das oder mit dem Medium (2) gelöst, gebunden, geschmolzen, gehärtet oder verklebt wird.

12. Verfahren nach einem der Ansprüche 1 bis 11, bei dem als Medium (2) ein flüssiges, gelartiges, thixotropes, pastöses, pulverförmiges, als Granulat vorliegendes oder festes Material verwendet wird, und/oder als Material (3) ein flüssiges, gelartiges, pastöses Material verwendet wird.

13. Verfahren nach einem der Ansprüche 1 bis 12, bei dem das Medium (2) aus der Gruppe ausgewählt ist, die Wasser, Gelatine, eine wässrige Polyaminlösung und eine Mischung davon enthält, und das Material (3) aus der Gruppe ausgewählt ist, die bei Raumtemperatur flüssige

Oligomere und Polymere, Schmelzen von Oligomeren und Polymeren, reaktive Oligomere und Polymere, Monomere, Gele, Pasten, Plastisole, Lösungen, Zwei-Komponenten-Systeme mit co reaktiven Komponenten, Dispersionen und Mischungen davon enthält, und/oder

14. Verfahren nach Anspruch 13, bei dem für das Material (3) als Gel ein oder mehrere ein- oder zwei-komponentige Silikonkautschuke, als Pasten ein oder mehrere gefüllte Oligomere und Polymere mit ein oder mehreren organischen und anorganischen Füllstoffen und als Zwei-Komponenten- Systeme mit co-reaktiven Komponenten ein oder mehrere Isocyanat/Polyamid-Systeme benutzt werden, oder als Material (3) ein oder mehrere Oligourethane verwendet werden.

15. Verfahren nach einem der Ansprüche 1 bis 14, bei dem anorganische und organische Füllstoffe im Medium (2) oder in dem Material (3) enthalten sind.

16. Verfahren nach einem der Ansprüche 1 bis 15, bei dem als Medium (2) ein oder mehrere Monomere verwendet werden, eine Faserstruktur und/oder eine Gerüststruktur in einer Matrix des Monomers oder der Monomere eingebaut wird und anschließend das oder die Monomere polymerisiert werden.

17. Verfahren nach einem der Ansprüche 1 bis 16, bei der die rheologischen Eigenschaften des Mediums (2) und des Materials (3) durch Verwenden organischer und anorganischer Nanofüllstoffe eingestellt werden.

18. Verfahren nach einem der Ansprüche 1 bis 17, bei dem biologisch aktive Substanzen in dem ersten und/oder in dem zweiten Material (2, 3) enthalten sind.

19. Verfahren nach Anspruch 18, bei dem ein oder mehrere Zelltypen an räumlich genau definierten Stellen zur Ausbildung einer genau einstellbaren dreidimensionalen Struktur abgegeben werden.

20. Verfahren nach Anspruch 19, bei dem Poren für die Nährstoffversorgung und für die Entsorgung von Stoffwechselprodukten in der dreidimensionalen Struktur vorgesehen werden.

21. Vorrichtung zur Ausführung des Verfahrens nach einem der vorhergehenden Ansprüche mit

einem Behälter (1) für das Medium (2),

einem dreidimensional bewegbaren Dispenser (4) zum Abge ben des Materials (3) in das Medium (2),

wobei der Dispenser (4) eine Ausgangsöffnung (5) auf weist, die unterhalb der Füllhöhe des ersten Materials (2) in dem Behälter (1) positionierbar ist.

22. Vorrichtung nach Anspruch 21, bei der die Ausgangsöffnung (5) als eindimensionale Düse oder als zweidimensionales Düsenfeld mit einzeln ansteuerba ren, individuell beheizbaren und/oder ventilgesteuerten Düsen ausgebildet ist,

und der Dispenser (4) ein oder mehrere Behälter für die Komponenten des Materials (3) aufweist.

23. Vorrichtung nach Anspruch 22, die so ausgebildet ist, daß das Medium (2) und/oder das Material (3) in einem definierten Zustand gehalten oder beim Abgeben gezielt eine thermisch induzierte Reaktion hervorgerufen wird, durch Heizen oder Kühlen der Behälter für die Komponen ten des Materials (3), und/oder des Behälters (1) und/oder der Düse(n).

24. Verwendung von biologisch oder pharmazeutisch aktiven Substanzen in einem Verfahren nach einem der Ansprüche 1 bis 20 und/oder einer Vorrichtung nach einem der Ansprüche 21 bis 23 zur Herstellung von biomedizinischen oder biologisch aktiven dreidimensionalen Objekten.

25. Verwendung nach Anspruch 24, wobei als biologisch oder pharmazeutisch aktive Substanzen Proteine, Wachstumsfaktoren und lebende Zellen, Hyaluronsäure, Gelatine, Collagen, Alginsäure und Ihre Salze, Chitosan und seine Salze als Zusätze oder als Matrixmaterial verwendet werden.

Andere Verfahren

3D Druck, mehrfarbig, mit verschiedenen Materialien

Veröffentlichungsnummer	US20020145213 A1
Publikationstyp	Anmeldung
Veröffentlichungsdatum	10. Okt. 2002
Eingetragen	10. Apr. 2001
Prioritätsdatum	10. Apr. 2001
Auch veröffentlicht unter	US6780368
Erfinder	Bor Jang, Junhai Liu
Ursprünglich Bevollmächtigter	Jang Bor Z., Junhai Liu

Referenziert von (18), Klassifizierungen (24), Legal Events (4)

Externe Links: USPTO, USPTO-Zuordnung, Espacenet

Kommentar:

Die Anmelder stellen ein Verfahren vor, das dem Prinzip des Laserdruckers ähnelt. Mittels elektrostatischer Aufladung wird ein Pulver ionisiert und auf eine Trägerplatte als Abbild einer Schichtlage des 3D Objekts erzeugt. Nicht ionisierte Stellen bleiben pulverförmig und ungebunden. Im nächsten Schritt wird nur das ionisierte Pulver mit einem Binder verfestigt oder auch entwickelt. Dann wird die Trägerplatte auf eine vorhergehende Lage Pulver lose transgerriert. Die Trägerplatte wird entladen und ist nun bereit für die Aufnahme von neuem Pulver um dieses erneut durch Ionisierung an sich zu binden. Die Schicht mit dem abgestreiften und gebunden Pulver wird in einem weiteren Schritt mit der darunterliegenden Schicht durch UV Aushärtung oder einem vergleichbaren Mittel ausgehärtet und verbunden. Theoretisch könnte man also durch verschiedene Pulver, verschiedene Farben und Materialien erzeugen. Das Pulverhandling in einer Maschine erscheint schwierig. Das ist aber bei einem Patent nicht relevant. Mit entsprechendem Aufwand ist es machbar, selbst wenn es diese heute noch nicht geben sollte. Obwohl hier gekürzt wiedergegeben, ist die Patentschrift sehr lang, aber daher auch ergiebig und sehr detailreich beschrieben. Für die vollständige Schrift bitte das Originaldokument aufrufen.

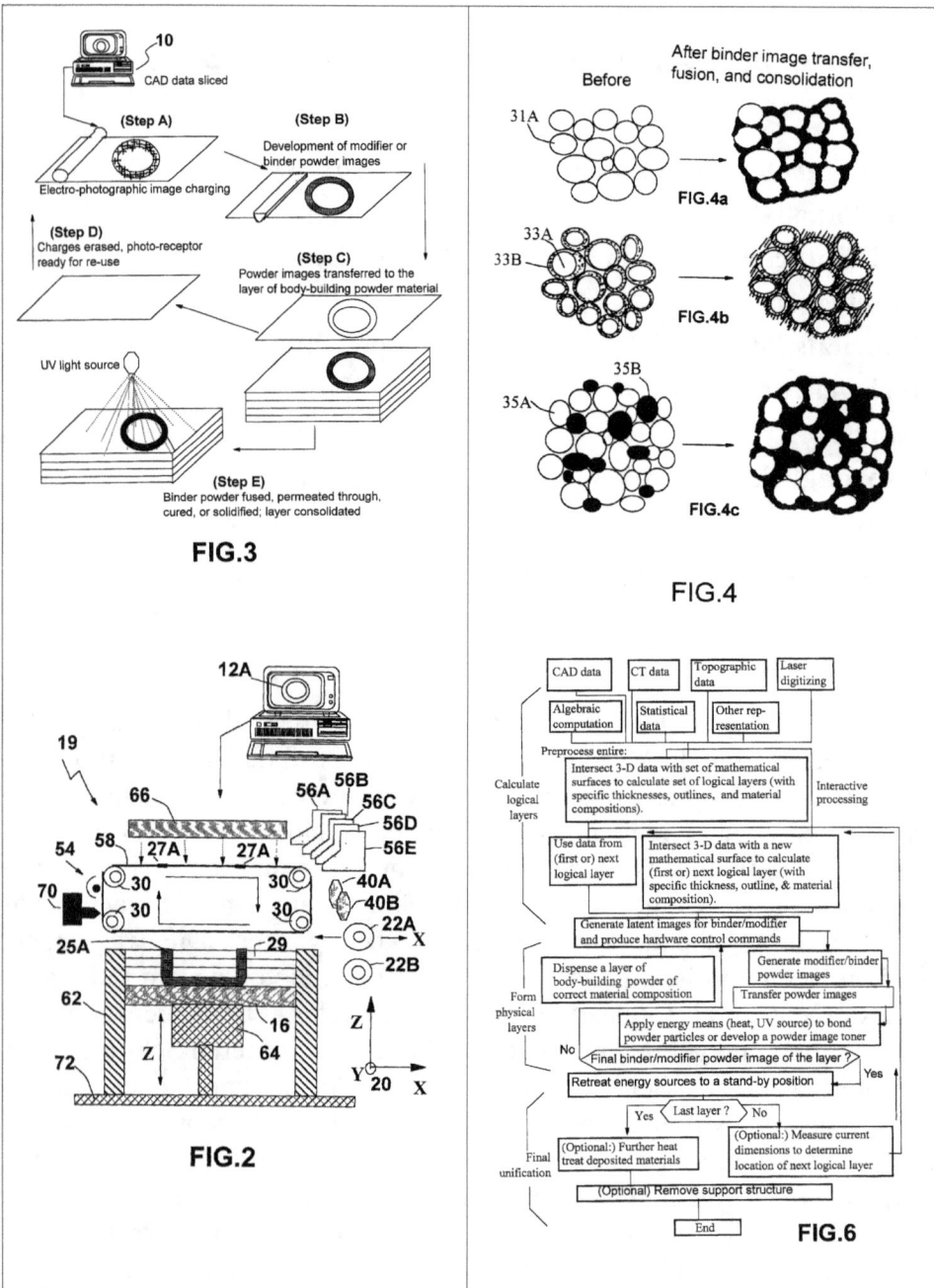

FIG.3

FIG.4

FIG.2

FIG.6

FIELD OF THE INVENTION

A solid freeform fabrication method and related apparatus for fabricating a three-dimensional, multi-material or multi-color object from successive layers of a primary body-building powder, at least a modifier powder and a binder powder in accordance with a computer-aided design of the object, the method including: (a) feeding a first layer of the primary body-building powder to a work surface; (b) operating an electrophotographic powder deposition device to create at least a modifier powder image and a binder powder image in accordance with this design; (c) transferring these powder images in a desired sequence to the first layer of a primary body-building powder; (d) applying energy sources to fuse the binder powder, forming a binder fluid that permeates through the first layer of a primary body-building powder for bonding and consolidating the powder particles to form a first cross-section of the object; (e) feeding a second layer of a primary body-building powder onto the first layer and repeating the operating, transferring, and applying steps to form a second cross-section (possibly of a different material composition distribution or color pattern) of the object; (f) repeating the feeding, operating, transferring, and applying steps to build successive layers of materials in a layer-wise fashion in accordance with the design for forming the multiple-layer, multi-material object; and (g) removing un-bonded powder particles, causing the 3-D object to appear.

BACKGROUND OF THE INVENTION

[0002] Layer manufacturing (LM) or solid freeform fabrication (SFF) or is a new fabrication technology that builds an object of any complex shape layer by layer or point by point without using a pre-shaped tool such as a die or mold. This process begins with creating a Computer Aided Design (CAD) file to represent the geometry or drawing of a desired object. This CAD file is converted to a proper solid interface format such as the stereo lithography (.STL) format. The geometry file is further sliced into a large number of thin cross-sectional layers with each layer being comprised of coordinate point data. In a commonly used layer-wise data format called Common Layer Interface (CLI), the contours (shape and dimensions) of each layer are defined by a plurality of line segments connected to form polylines on an X-Y plane of a X-Y-Z orthogonal coordinate system. The layer data are converted to tool path data normally in terms of computer numerical control (CNC) codes such as G-codes and M-codes. These codes are then utilized to drive a fabrication tool for defining the desired areas of individual layers and stacking up the object layer by layer along the Z-direction.

(..)

[0003] The SFF technology makes it possible to convert a CAD image data directly into a three-dimensional (3-D) physical object. The technology has been widely used in applications such as verifying CAD database, evaluating engineering design feasibility, testing part functionality, assessing aesthetics, checking ergonomics of design, aiding in tool and fixture design, creating conceptual models and marketing tools, producing medical or dental models, generating patterns for investment casting, reducing or eliminating engineering changes in production, and providing small production runs.

[0004] The SFF techniques may be divided into three categories: layer-additive, layer-subtractive, and hybrid (combined layer-additive and subtractive). A layer additive process involves adding or depositing a material to form predetermined areas of a layer essentially point by point; but a multiplicity of points may be deposited at the same time in some techniques, such as of the multiple-nozzle inkjet-printing type. These predetermined areas together constitute a thin cross-section of a 3-D object as defined by a CAD geometry. Successive layers are then deposited in a predetermined sequence with a layer being affixed to its adjacent layers for forming an integral multi-layer object. A 3-D object, when sliced into a plurality of constituent layers or thin sections, may contain features that are not self-supporting and in need of a support structure during the object-building procedure. These features include isolated islands in a layer and overhangs. In these situations, additional steps of building the support structure, also on a layer-by-layer basis, will be required of a layer-additive technique. An example of a layer-additive technique that normally requires building a support structure is the fused deposition modeling (FDM) process as specified in U.S. Pat. No. 5,121,329; issued on Jun. 9, 1992 to S. S. Crump.

[0005] A layer-subtractive process involves feeding a complete solid layer of a material to the surface of a support platform and using a cutting tool (normally a laser) to cut off or somehow degrade the integrity of the un-wanted areas of this solid layer. The solid material in these un-wanted areas of a layer becomes a part of the support structure for subsequent layers. These un-wanted areas are hereinafter referred to as the "negative region" while the remaining areas that constitute a cross-section of a 3-D object are referred to as the "positive region". A second solid layer of material is then fed onto the first layer and bonded thereto. The same cutting tool is then used to cut off or degrade the material in the negative region of this second layer. These procedures are repeated successively until multiple layers are laminated to form a unitary object. After all layers have been completed, the unitary body (or part block) is removed from the platform, and the excess material (in the negative region) is removed to reveal the 3-D object. This "decubing" procedure is known to be tedious and difficult to accomplish without damaging the object. An example of a layer-subtractive technique is the well-known laminated object manufacturing (LOM), disclosed in, for instance, U.S. Pat. No. 4,752,352 (Jun. 21, 1988 to M. Feygin).

[0006] A hybrid process involves both layer-additive and subtractive procedures. An example can be found with the Shape Deposition Manufacturing (SDM) process disclosed in U.S. Pat. No. 5,301,863 issued on Apr. 12, 1994 to Prinz and Weiss. Such a process is complicated and difficult to operate. It also requires the operation of heavy and expensive equipment.

[0007] Another good example of the layer-additive technique is the 3-D powder printing technique (3D-P) developed at MIT; e.g., U.S. Pat. No. 5,204,055 (April 1993 to Sachs, et al.) and U.S. Pat. No. 6,007,318 (Dec. 28, 1999 to Russell, et al.). This 3-D powder printing technique involves dispensing a layer of loose powders onto a support platform and using an ink jet to spray a computer-defined pattern of liquid binder onto a layer of uniform-composition powder in a point-by-point fashion. The binder serves to bond together the powder particles on those areas (positive region) defined by this pattern. Those powder particles in the un-wanted areas (negative region) remain loose or separated from one another and are removed at the end of the build process. Another layer of powder is spread over the preceding one, and the process is repeated. The "green" part made up of those bonded powder particles is separated from the loose powders when the process is completed. This procedure is followed by binder removal and impregnation of the green part with a liquid material such as epoxy resin and metal melt. Although several nozzle orifices may be employed to dispense several droplet streams at the same time, this 3D-P process remains to be essentially a point-by-point process, being characterized by a slow build speed.

[0008] This same drawback is true of the selected laser sintering (SLS) technique (e.g., U.S. Pat. No. 4,863,538, Sep. 5, 1989 to C. Deckard, U.S. Pat. No. 4,938,816, Jul. 3, 1990 to J. Beaman, et al., and U.S. Pat. No. 5,316,580, May 31, 1994 to Deckard). The SLS technique involves spreading a full-layer of loose powder particles and uses a computer-controlled, high-power laser to partially melt these particles within predetermined areas (positive region) in a point-by-point fashion. Commonly used powders include thermoplastic particles, thermoplastic-coated metal particles, metal-coated ceramic particles, and mixtures of high-melting and low-melting powder materials. These point-wise procedures are repeated for subsequent layers, one layer at a time, according to the CAD data of the sliced-part geometry. The loose powder particles in the negative region of each layer are allowed to stay as part of a support structure. The sintering process does not always fully melt the powder, but allows molten material to bridge between particles. Commercially available systems based on SLS are known to have several drawbacks. One problem is that the need to use a high power laser makes the SLS an expensive technique and un-suitable for use in an office environment. Again, the spot-by-spot or point-by-point laser scanning is a very slow procedure, resulting in a low object-building speed.

[0009] In U.S. Pat. No. 5,514,232, issued May 7, 1996, Burns discloses a method and apparatus for automatic fabrication of a 3-D object from individual layers of fabrication material having a predetermined configuration. Each layer of fabrication material with desired shape and dimensions is first deposited on a carrier substrate in a deposition station. The fabrication material along with the substrate are then transferred to a stacker station. At this stacker station the individual layers are stacked together, with successive layers being affixed to each other and the substrate being removed after affixation. Lamination-based LM techniques that require radiation curing of solid sheet polymer materials layer by layer can be found in U.S. Pat. No. 5,174,843 (Dec. 29, 1992 to M. Natter) and No. 5,352,310 (Oct. 4, 1994 to M. Natter). Natter's technique is limited to high-energy radiation-curable polymer materials in a solid sheet form. Disclosed in U.S. Pat. No. 5,183,598 (Feb. 2, 1993 to J-L Helle, et al.) is a process that includes preparing thin sheets of a fiber- or screen-reinforced matrix material. In these composite sheets, the matrix material exhibits the feature that its solubility in a specific solvent can be changed when the material is exposed to a specific radiation. Selected areas of individual sheets are radiated to reduce the solubility. The un-irradiated portion (the negative region) of individual layers remains soluble in the solvent. The stack of sheets are affixed together to form an integral body, which is immersed in the solvent that causes the desired object to appear. This process exhibits the following shortcomings:

[0010] (1). A high-power radiation source (e.g., a high-power laser beam) is required. High energy radiation sources and their handling equipment (for reflecting, focusing, etc) are expensive. Furthermore, they are not welcome in an office environment.

[0011] (2). When a screen is used as the reinforcement, the screen in the negative region is difficult to get dissolved in the solvent particularly if this screen is made of metal or ceramic materials. A strong acid is needed in dissolving a metal screen.

[0012] Lamination-based LM techniques that involve transferring thin sections of powders, prepared by electrophotographic or electrostatic attraction, to a stacking station are disclosed in U.S. Pat. No. 5,088,047 (Feb. 11, 1992 to D. Bynum), U.S. Pat. No. 5,593,531 (Jan. 14, 1997 to S. M. Penn), and U.S. Pat. No. 6,066,285 (May 23, 2000 to Kumar). In Bynum's process, a drum-shaped electrophotographic element is first prepared. A light image corresponding to a cross-section of an object generated by a computer is projected into this element by line-by-line laser scanning, coordinated with rotational speed of the drum to selectively dissipate the charge thereon, thereby creating an electrostatic latent image on the element. The element, along with the latent image thereon, is then rotatably transferred to a plurality of developer stations, which respectively apply forming powders (toner) to different areas of the electro-photographic element. For each layer, at least two developer stations are needed to apply two different powders to the positive and negative regions, respectively, for building the object cross-section (positive region) and the support structure (negative region). These areas of powders are then electrostatically attracted to a surface of an endless flexible belt, which carries these patterned powders to a fixing station where the powder particles in the positive region are made tacky by the application of heat or solvent vapor. The tackified lamina is then transferred to a stacking station and laid up onto a support platform or a previous layer to form a layer of both the object cross-section and support structure. The above steps are repeated in the same sequence to lay up multiple laminas to form a block of laminas. The powder materials in the negative regions for forming the support structure are usually made of lower melting materials and can be removed by heat from this block at the end of the build process to reveal the desired 3-D object. A fundamentally similar process is disclosed in Penn's patent and Kumar's patent. The processes specified in these three patents (U.S. Pat. Nos. 5,088,047, 5,593,531, and 6,066,285) have the following drawbacks:

[0013] (1) At least two toner developing stations are required, one for forming the part (object) and the other for the support structure. For every layer of the same object-building material, two different types of powders have to be precisely deposited electrostatically, in sequence and in registration, onto complementary areas of a layer. This is difficult to accomplish without suffering cross-contamination.

[0014] (2) It is well-known in the art of electrophotography that most of the conductive particles (e.g., metal powders) do not work well with charging devices.

This effectively eliminates the freeform formation of many metallic parts if metal particles are the primary body-building material of the part being built. In contrast, the presently invented method provides an effective way of eliminating this limitation, making our method so much more versatile. In this method, we make use of a simple powder-feeder to supply and evenly spread up a layer of a primary object body-building powder material (e.g., metal), analogous to the powder-feeding step in afore-mentioned SLS and 3D-P processes. We then use electrophotography techniques to form, develop, and transfer toner images of a binder powder (to bond or sinter together the underlying primary body-building powder particles) and a plurality of property modifying powders (modifier powders, e.g., coloring agent), simultaneously or in sequence. The binder and modifier powders collectively occupy only a small fraction of the object cross-section being built.

[0015] (3) These three prior art electrophotography methods are limited to loose powders as the starting primary body-building materials. Other forms of material such as a porous substrate (e.g., comprising fiber preform as a reinforcement for a composite) can not be used in these processes.

[0016] (4) Penn's and Kumar's methods are essentially limited to the fabrication of an object of homogeneous material composition and are not easily or readily adapted for the preparation of a multi-material or multi-color object in which the material composition or color pattern can be varied from point to point. Bynum's method, in principle, allows for variation of material composition or color pattern from point to point, like in the case of the traditional 2-D printing process that involves developing and transferring multi-color toner images to a sheet of paper. In real practice, however, the electrostatic attraction in a traditional electrophotography system can only handle a thin layer of light-weight toner powder at a time, up to 10 μm or less in thickness. It would take an extremely long time to build up a 3-D model of, say, 100 mm in thickness. In contrast, in our method, the powder feeder can feed layers of heavy- or light-weight powder of which the layer thickness can be varied from very thin to very thick. With the primary body-building powder occupying the majority of the object volume (typically 70% to 95%), the electrophotography device is required to provide only a small amount of binder and modifier powders at a time. Further, in our method, in the negative region of a layer where the primary body-building powder receives no binder, the powder particles serve to provide the needed support structure. It is not necessary to carry out the extra steps of developing a support structure toner image and transferring this image to the negative region of a layer (where the positive region of the layer is already deposited with the image material) in such a fashion that the two complementary regions of different materials must perfectly match (in registry) in shapes and thickness.

[0017] Despite these shortcomings of the afore-mentioned three patents, the concept of adapting electrophotography techniques for transferring powder materials in a LM system has proven to be very useful.

[0018] Due to the specific solidification mechanisms employed, many LM techniques are limited to producing parts from specific polymers. For instance, Stereo Lithography (SLa) and Solid Ground Curing (SGC) rely on ultraviolet (UV) light induced curing of photocurable polymers such as acrylate and epoxy resins. The photo-curable polymer in these two cases constitutes the vast majority of the material in the resulting 3-D object. Any other ingredient such as an additive or reinforcement represents at best a minority phase in the structure. The photo-curable polymer in the resulting structure is a "host" while any additive, if present, is just a guest. The host provides the basic structural integrity of the 3-D object. Unfortunately, photo-curable polymers alone normally do not have good mechanical strength and toughness.

[0019] The above state-of-the-art review has indicated that all prior-art layer manufacturing techniques have serious drawbacks that prevent them from being more widely implemented.

[0020] Therefore, an object of the present invention is to provide an improved layer-additive method and apparatus that can be used for producing a multi-material or multi-color 3-D object.

[0021] Another object of the present invention is to provide a computer-controlled method and apparatus for producing a part on a layer-by-layer, but not point-by-point basis (hence, with a high build speed).

[0022] It is a further object of this invention to provide a computer-controlled object building method that does not require heavy and expensive equipment such as a high-power laser system.

[0023] It is another object of this invention to provide a method and apparatus for building a CAD-defined object in which the support structure is readily provided during the layer-adding procedure.

[0024] Still another object of this invention is to provide a layer manufacturing technique that places minimal constraint on the range of materials that can be used in the fabrication of a 3-D object. Further, the material composition or color of the object can be varied from spot to spot and/or from layer to layer.

SUMMARY OF THE INVENTION

[0025] The Method

[0026] The objects of the invention are realized by a method and related apparatus for fabricating a three-dimensional, multi-material or multi-color object on a layer-by-layer basis (but not point-by-point) and in accordance with a computer-aided design (CAD) of this object. The object is made from at least a primary body-building powder material, a binder powder, and at least a property-modifying material in fine powder form (hereinafter referred to as modifier powder). This modifier powder can contain a colorant. The design contains data on the geometry (shape and dimensions) and material composition distribution (and/or color pattern). The data preferably is sliced into layer-wise data sets with each set defining the geometry and material composition of a constituent cross-section of the object. Basically, the method includes, in combination, the following steps:

[0027] (a) providing a work surface on a support platform that lies substantially parallel to an X-Y plane of an X-Y-Z Cartesian coordinate system defined by three mutually orthogonal X-, Y- and Z-axes;

[0028] (b) feeding a first layer of a primary body-building powder material to the work surface (e.g., by using a traditional powder feeder commonly used in selected area sintering and 3-D powder printing processes);

[0029] (c) operating an electrophotographic powder deposition means to create transferable powder toner images of a binder powder and at least a modifier powder in accordance with the CAD design; (A plurality of modifier powders may form separate toner images or may be combined to form one composite toner image.)

[0030] (d) transferring the transferable modifier and binder powder images, one image at a time, in a desired sequence onto the first layer of the primary body-building powder material;

[0031] (e) applying energy means to fuse said binder powder, allowing the resulting fused binder fluid to permeate downward through the first layer of primary body-building material for bonding and consolidating the particles in the first layer to form a first cross-section of the object; (Bonding and consolidating are hereinafter collectively referred to as sintering in the present context.)

[0032] (f) feeding a second layer of a primary body-building powder material onto the deposited first layer and repeating the operating, transferring, and applying steps to form a second cross-section of the object; (The material distribution and color pattern in the second cross-section may be different from those of the first cross-section.)

[0033] (g) repeating the feeding, operating, transferring, and applying steps to build successive layers of possibly varying material compositions and/or color patterns in a layer-wise fashion in accordance with the CAD design for forming multiple layers of the object; and

[0034] (h) removing un-bonded powder particles, causing the 3-D object to appear.

[0035] In this instant method, the steps of applying energy means could include pre-heating a layer of primary body-building powder material to a temperature above the melting point of the binder powder. This is done so that the binder powder, when transferred and deposited onto the predetermined areas (positive region) of a corresponding pre-heated body-building material powder layer, will be quickly melted to become a fluid that permeates through the gaps between fine particles of the body-building material powder. This binder fluid, when solidified, will bond and consolidate the powder particles in the positive region, leaving the powder particles in the negative region un-bonded (free from binder). The particles in the negative region stay as part of a support structure. As one can easily see, in this method, any material that can be made into a fine powder form can be used as a primary body-building material and can be easily fed and evenly spread up to form a layer. This is a very significant advantage over other prior art electrophotography-based LM techniques.

[0036] The binder powder could include a resin composition that can be cured or hardened with heat, ultra violet light, electron beam, ion beam, plasma, microwave, X-ray, Gamma ray, or a combination thereof. Alternatively, the binder powder could include a lower-melting material that can be readily fused to become a fluid. Once permeating through a layer of primary body-building powder material for providing bridges between particles, the binder fluid can be cooled down to below the melting point of the binder material and be solidified. Preferably, the steps of applying energy means are carried out in such a manner that successive layers are affixed together to form a unitary body of the 3-D object. This can be easily accomplished by allowing the fused binder fluid to have sufficient time to permeate through the current layer of body-building powder material and reaching the top surface of the previously deposited layer.

[0037] In the instant invention, the working principle of the electrophotographic powder deposition means can be selected from a range of electrostatic printer or photocopier mechanisms. For instance, electrophotographic powder deposition means can include, but not limited to (1) planar capacitor dot matrix charging device and (2) combined corona discharging/thin photoconductive charge receptor/scanning laser imaging devices. The electrophotographic powder deposition means is characterized by the following features:

[0038] (4) It provides a 2-D pattern or "latent image" of electrostatic charges to attract fine powder particles of the binder composition and/or modifiers to form these binder/modifier particles into a toner "image" (thin section of powder particles) in selected areas of a powder layer; these areas being programmable and predetermined by a computer. These areas, corresponding to the positive region of a layer, are defined by the layer data of a CAD design for the object to be built. A full area of the binder powder and/or modifier powder is formed and transferred to deposit onto a layer of body-building powder material, equivalent to a process of "photo-printing". The binder powder "photo-printed" to the positive region of a body-building powder material layer will help sinter the particles therein, forming a cross-section of the 3-D object. The modifier powder image transferred to the same region of a layer will impart desired physical properties (e.g., color appearance) to this layer. The primary body-building powder particles in other areas of the same layer, not receiving any binder powder composition, will remain as isolated, loose particles that serve as part of a support structure. As opposed to the case of conventional selected laser sintering (SLS) in which a laser beam is used to sinter the powder spot by spot (essentially point by point), the presently invented method builds the part area by area (up to one full layer at a time). This is also in sharp contrast to operating an inkjet printhead to print adhesive onto a layer of powder in a point-by-point fashion in a conventional 3D powder printing (3D-P or MIT) process.

[0039] (5) The binder powder, once deposited, is melted in such a manner that the binder fluid flows around to provide a bridge between primary body-building particles in the positive region. The binder can bond together these particles to impart sufficient strength and rigidity to the layer for easy handling and for maintaining the part dimensional accuracy during the formation of subsequent layers. If the binder contains a photo-curable adhesive composition, the pre-heat energy intensity and the energy of the imposing light source (heat and light constituting the energy means) should be provided in such a fashion that successive layers can be affixed together to form a unitary body of the 3-D object.

[0040] (6) If the binder contains a heat-fusible material composition, a complete body-building powder layer can be pre-heated by other heat sources (e.g., infrared, IR) disposed near the object-building zone to a temperature (Tpre) sufficient for melting the binder composition. After a selected duration of time, this heat source may be switched off to allow the binder fluid (already permeating through a layer) to solidify. If the layer of primary body-building material is already mixed with component compositions of a binder (excluding a photo-initiator, for instance), the electro-photographic powder deposition means may be used to transfer an image of the photo-initiator powder to the positive region of the layer. The pre-heat temperature Tpre may be so chosen that it is capable of promoting the curing reaction once initiated by the photo-initiator along with an incident light, but insufficient for initiating the curing reaction of the binder compositions by the pre-heat alone. This auxiliary heat would help accelerate the cure reaction and significantly reduce the light intensity requirement that would otherwise be imposed upon the light source. In this favorable situation, the light source can be just based on an ordinary ultraviolet (UV) light source. No expensive high-power laser beam, electron beam, X-ray, Gamma-ray or other high-energy radiation is necessary.

[0041] (7) The physical sizes of the binder powder image forming area (electrostatically charged substrate area of a photo-receptor, for instance) of this electrophotographic powder deposition means are preferably sufficient to cover the complete envelop of a primary body-building powder layer so that a complete cross-section of the 3-D object can be built in one binder powder image transfer. This is one of the advantages over the case of conventional selected laser sintering (SLS) which requires aiming a laser beam to one spot at a time (spot being micron- or sub-millimeter-sized). It would take a much longer time for a laser beam to fuse and sinter the particles of a complete cross-section in a spot-by-spot or point-by-point fashion. Further, since binder powder image can be exactly identical to the desired cross-section of a layer, this instant invention also has a significant advantage over the

conventional 3D-P process, which involves ejecting adhesive droplets essentially point by point to cover the positive region, a slow process indeed.

[0042] In the presently invented method, the photo-curable binder may consist of such adhesive compositions as a base resin, a hardening or cross-linking agent, a photo-initiator, a photo-sensitizer, and possibly with additional catalyst and/or reaction accelerator. All of these compositions, if in a powder form, may be mixed together to form a complete binder adhesive mixture. This binder mixture is then attracted by the electro-photographic means to form into a binder image, which is transferred and deposited onto a powder layer. Alternatively, one or more compositions may be included as secondary ingredients in the primary body-building powder material to be dispensed one layer at a time by a powder feeder (powder-dispensing means) while the remaining composition(s) may constitute the binder powder image.

[0043] The powder inside a powder feeder may comprise a primary body-building material (fine particles), additives (physical or chemical property modifiers), and secondary ingredients (selected compositions of a binder adhesive). In this method, the primary body-building powder may be composed of one or more than one type of fine particles. These fine powder particles could be of any geometric shape, but preferably spherical. The particle sizes are preferably smaller than 100 μm, further preferably smaller than 10 μm, and most preferably smaller than 1 μm. The size distribution is preferably uniform. The primary body-building powder may be selected from the following three basic types of powders:

[0044] Type A: fine particles of a primary

body-building material only. In this type, only primary body-building materials in a fine particle form are included as the ingredients in the powder; no binder composition being included. All binder compositions are present as a binder powder to be formed into an image by the electro-photographic means. The primary body-building materials can be selected from polymers, ceramics, glass, metals and alloys, carbon, and combinations thereof. The polymers may be thermoplastic (e.g., polyvinyl chloride) or thermosetting (e.g., polyimide oligomer or prepolymer powder). The binder, including all selected compositions, will be deposited over the positive region of a complete layer and allowed to permeate through the gaps between fine particles in a layer of primary body-building powder. The binder (if an adhesive) in the positive region (corresponding to the desired cross-section) of a layer will be at least partially cured (chemically cross-linked or otherwise hardened) to bond together the primary body building particles. The binder (if containing a fusible material composition) will be heated to become a fluid which, once permeated through a layer, will be cooled to solidify. No binder will be deposited to the negative region and, hence, the fine particles in this region will remain loose and will stay as part of a support structure.

[0045] Type B: fine ceramic, metallic, glass, or polymeric particles (as primary body-building materials) each coated with a thin layer of coating comprising selected binder adhesive compositions. Once a layer of these coated solid particles is deposited, the remaining compositions of a binder adhesive are then deposited, melted, and allowed to permeate through the gaps between these primary body-building particles. These remaining compositions are then in contact or reacted with the selected binder compositions in the coating to make a complete binder adhesive. The binder adhesive, only existing in the positive region of a layer, is then at least partially cured by heat and/or UV light or any other energy means to bond together body-building particles, leaving the particles in the negative region loose and un-bonded.

[0046] Type C: a mixture of fine particles of primary body-building materials (e.g., a silicon dioxide powder) with at least one binder adhesive composition also in a fine powder form. The other remaining binder adhesive compositions are electro-photographically formed into a binder image, deposited onto a layer of Type C powder mixture, and allowed to flow around the fine particles and react with the at least one binder adhesive composition. The complete binder adhesive formulation in the positive region of this layer is then at least partially cured to provide inter-particle bonding for those primary body-building particles in the positive region. Again, the adhesive will not enter the negative region and the powder particles in this region will remain loose and physically separable.

[0047] In each powder type, additional ingredients may be added to impart desired physical and/or chemical properties to the object being built. These ingredients may contain a reinforcement composition selected from the group consisting of short fiber, whisker, and particulate reinforcements such as a spherical particle, ellipsoidal particle, flake, small platelet, small disc, etc. These ingredients may also contain, but not limited to, colorants, anti-oxidants, anti-corrosion agent, sintering agent, plasticizers, etc. Any of these ingredients, when intended to be used in each and every layer of the 3-D object, may preferably be included in the primary body-building powder to be dispensed by a traditional powder feeder. Those ingredients that are to be deposited only at selected spots of a layer or selected layers (but not all layers) of an object may be included as a part of a modifier powder. These ingredients will then be electrophotographically formed into a modifier powder image (toner) and transferred to a corresponding cross-section of a primary body-building powder, before or after the binder powder image is transferred. Alternatively, selected ingredients may be combined with a binder powder to form a composite binder-modifier powder image.

[0048] Many prior-art powder-dispensing means or feeders are available for feeding layers of powder materials, one layer at a time. The moving and dispensing operations of the powder-dispensing means and the operation of an electrophotographic powder deposition means are preferably conducted under the control of a computer. This can be accomplished by (1) first creating a computer-aided design of the 3-D object on a computer with the design containing information on both the geometry and material composition distribution of the object with the geometry including a plurality of data points defining the object, (2) generating programmed signals corresponding to each of the data points, collected into layer-wise data sets, in a predetermined sequence; (3) generating plural powder images (comprising a binder powder image and at least a modifier powder image) and transferring/depositing these binder/modifier powder images to corresponding areas of a layer of body-building powder material responsive to these programmed signals, (4) moving the powder-dispensing means and the work surface relative to each other (in Z-direction, e.g.) in response to these programmed signals. The signals for moving may be advantageously prescribed in accordance with the G-codes and M-codes that are commonly used in computer numerical control (CNC) machinery industry, but other motion control codes may also be used. The signals for forming a powder image may be created by any image formation means commonly used in an electrostatic printer or photo-copier.

[0049] In order to produce a multi-material 3-D object in which the material composition of the primary body-building powder can vary from layer to layer, the presently invented method may further comprise the steps of (1) creating a geometry of the 3-D object on a computer with the geometry including a plurality of layer-wise sets of data points defining the object; each of the data sets being coded with a selected material composition, (2) generating programmed signals corresponding to each of the data sets in a predetermined sequence; and (3) operating the powder-dispensing means in response to the programmed signals to dispense and deposit powders of selected body-building material compositions, with the material compositions varying possibly from layer to layer. In order to achieve a point-to-point variation in material composition or color, each data point may be coded with a material composition or color. Such a material distribution or color pattern can be physically achieved by using the color electrophotography steps to form and transfer multi-material or multi-color powder images to corresponding layers of a primary body-building powder. The virtual reality modeling language (VRML), which is capable of building the geometry of a 3-D object with rich material composition and/or color information, is particularly useful as a CAD tool in the practice of the present invention.

[0050] To further ensure the part accuracy and compensate for the potential variations in part dimensions (thickness, in particular), the present method may be executed under the assistance of dimension sensors. These sensors may be used to periodically measure the dimensions of the object being built while a computer is used to determine the thickness and outline of individual layers intermittently in accordance with a computer aided design representation of the object. The computing step includes operating the computer to calculate a first set of logical layers with specific thickness and outline for each layer and then periodically re-calculate another set of logical layers after periodically comparing the dimension data acquired by the sensor with the computer aided design representation in an adaptive manner.

[0051] The Apparatus

[0052] Another embodiment of this invention is a solid freeform fabrication apparatus for automated fabrication of a 3-D object. This apparatus includes:

[0053] (1) a work surface to support the object while being built;

[0054] (2) powder-dispensing means at a predetermined initial distance from the work surface; the dispensing means having an outlet directed to the work surface for feeding successive layers of powder onto the work surface, one layer at a time, with the powder including at least a primary body-building material;

[0055] (3) an electrophotographic powder deposition means at a distance from the work surface; the electrophotographic powder deposition means having an imaging surface directed to the work surface for feeding successive layers of binder/modifier powder images onto the corresponding layers of primary body-building materials, one layer at a time;

[0056] (4) energy means at a distance from the work surface for providing fusion, cooling, curing, and/or bonding energy to successive layers being built; and

[0057] (5) motion devices coupled to the work surface, electrophotographic powder deposition means, and powder-dispensing means for moving the electrophotographic and dispensing means with respect to the work surface so that the binder/modifier powder image plane is substantially parallel to a plane defined by first and second directions (X- and Y-directions) and in a third direction (Z-direction) orthogonal to the X-Y plane to dispense multiple layers of powder and then transferring binder/modifier powder images, one layer at a time, for forming the 3-D object. Preferably, the work surface is lowered by one layer thickness distance vertically in the Z-direction after one layer is built to get ready for receiving powders of the next layer.

[0058] In order to automate the object-fabricating process, the present apparatus is preferably equipped with a computer-aided design computer and supporting software programs operative to (a) create a three-dimensional geometry of the 3-D object, (b) convert this geometry into a plurality of data points defining geometry and material composition of the object, and (c) generate programmed signals corresponding to each of the data points in a predetermined sequence. The apparatus also includes a three-dimensional motion controller electronically linked to the computer and the motion devices. The electrophotographic powder deposition means is also preferably electronically connected to the computer, optionally through an electrophotography controller. The motion controller is operated to actuate the motion devices and the electrophotography controller is operated to activate the electrophotographic powder deposition means to generate a binder and/or modifier powder image, both being responsive to the programmed signals for the data points received from the computer.

[0059] The apparatus preferably includes dimension sensors that are electronically linked to the computer. The sensors periodically provide layer dimension data to the computer. In the mean time, the supporting software programs in the computer act to perform adaptive layer slicing to periodically create a new set of layer data, including the data points defining the object, in accordance with the layer dimension data acquired by the sensors means. New sets of programmed signals corresponding to each of the new data points are generated in a predetermined sequence.

[0060] Specifically, the motion devices are responsive to a CAD-defined data file which is created to represent the 3-D preform shape to be built. A geometry (drawing) of the object is first created in a CAD computer. The geometry is then sectioned into a desired number of layers with each layer being comprised of a plurality of data points. These layer data are then converted to form an image for attracting binder powder particles and also converted to machine control languages that can be used to drive the operation of the motion devices and powder-dispensing devices. These motion devices operate to provide relative rotational and translational motions of the powder-dispensing device and the electro-photographic powder deposition means with respect to the work surface. The motion devices further provide relative movements of the work surface in the Z-direction, each time by a predetermined thickness distance.

ADVANTAGES OF THE INVENTION

[0061] The process and apparatus of this invention have several features, no single one of which is solely responsible for its desirable attributes. Without limiting the scope of this invention as expressed by the claims which follow, its more prominent features will now be discussed briefly. After considering this brief discussion, and particularly after reading the section entitled "DESCRIPTION OF THE PREFERRED EMBODIMENTS" one will understand how the features of this invention offer its advantages, which include:

[0062] (1) The present invention provides a unique and novel method for producing a three-dimensional object on a layer-by-layer basis under the control of a computer. This method does not require the utilization of a pre-shaped mold or tooling.

[0063] (2) Most of the layer manufacturing methods, including powder-based techniques such as 3-D printing (3DP) and conventional selective laser sintering (SLS), are normally limited to the fabrication of an object in a point-by-point fashion and, hence, are very slow. In contrast, the presently invented method allows the fabrication of a part one complete layer at a time due to the full-field sized programmable, electrophotographic powder deposition device being capable of precisely forming a thin layer of binder powder corresponding to the positive region of a

layer. Therefore, the presently invented method can be order-of-magnitude faster than the conventional SLS and 3DP.

[0064] (3) The presently invented method provides a computer-controlled process which places minimal constraint on the variety of materials that can be processed. In the present method, both the primary body-building powder material and the modifier powder may be selected from a broad array of materials including various organic (including polymers) and inorganic substances (including ceramic, metal, glass, and carbon based materials) and their mixtures. This is in sharp contrast to both Stereo Lithography (SLa) and Solid Ground Curing (SGC), which solely rely on ultraviolet (UV) light-curable polymers such as acrylate and epoxy resins as the primary body-building material. The photo-curable polymer in both SGC and SLa represents the vast majority of the material in the resulting 3-D structure and is the "matrix" or "host" that accommodates any additive or reinforcement that might exist in the structure. The host basically provides the structural integrity of the 3-D object. The cured resin will not be removed or otherwise disintegrated. In the instant invention, the binder adhesive provides only a vehicle for tentatively holding together other otherwise loose powder particles. This binder or adhesive constitutes only a minority material phase of the resulting 3-D structure. In the cases of ceramic, glass, or metal powder particles, this cured adhesive will be burned off leading to the formation of a somewhat porous structure. This porous structure is then either sintered at a high temperature to produce a solid body or impregnated with another liquid material (e.g., metal melt) to form a composite or hybrid material object. This final structure will contain no low-temperature material

such as the polymeric adhesive (only metal and/or ceramic, e.g.). Both metal and ceramic materials can be used in a much higher temperature environment.

[0065] In terms of the variety of materials, the presently invented method also presents several advantages over the prior-art electrophotographic powder deposition based SFF techniques. For instance, these prior-art techniques are normally limited to the formation of thin, light weight powder images only and are not able to form a thicker layer of heavier powders such as ceramic and metallic particles due to the limited electrostatic attractive force between charges and solid powder particles. Further, it is normally very difficult to charge electrically conductive materials such as metals and, hence, the prior-art electro-photographic methods are not effective in building parts from metallic powders. In contrast, in the practice of our method, one is free to choose any light-weight, non-conductive binder powder composition to be electrophotographically formed and transferred to a layer of primary body-building powder. Individual layers of a heavier and/or conductive primary body-building powder such as a metal or ceramic material can be deposited by using other more simple and easy-to-perform powder-dispensing means (such as those successfully used in SLS and 3D-P), which are not limited by the relatively weak electrostatic attractive forces.

[0066] (4) The present method provides an adaptive layer-slicing approach and a thickness sensor to allow for in-process correction of any layer thickness variation. The present invention, therefore, offers a preferred method of layer manufacturing when part accuracy is a desirable feature.

[0067] (5) The method can be embodied

using simple, inexpensive, and field-proven photo-copier mechanisms, so that the fabricator apparatus can be relatively small, light, inexpensive and easy to maintain. No high-power laser beam (to fuse and sinter a thicker layer of powder) is required.

[0068] (6) In the present method, a support structure naturally exists when a layer of body-building powder is fed. No additional tool is needed to build the support structure. This is in contrast to most of the prior-art layer-additive techniques that require a separate tool to build a support structure point by point, thereby slowing down the part-building process.

DESCRIPTION OF THE PREFERRED EMBODIMENTS

[0075] In the drawings, like parts have been endowed with the same numerical references. FIG. 1 illustrates one preferred embodiment of the presently invented apparatus for making a three-dimensional object. This apparatus is equipped with a computer 10 for creating a CAD drawing 12 (geometry and color pattern) of an object (shown as a coffee cup) and, through a hardware controller 14 (including signal generator, amplifier, and other needed functional parts) for controlling the operation of other components of the apparatus. These other components include at least a powder-dispensing means 22, an electrophotographic powder deposition means (of which a photo-receptor 18 and a binder powder image 27 being shown in FIG. 1), an energy means (UV source 40, as an example), and a work surface 16 on an object-supporting platform. The hardware controller 14 may comprise a UV light source controller, electrophotographic device controller, powder-dispensing device controller, and

motion controller. The powder-dispensing means 22 provides successive layers of a primary body-building powder material onto the work surface 16 one layer at a time. A plurality of powder-dispensing means (one of the powder feeders being shown as 22 in FIG. 1) may be used to feed successive layers of different primary body-building powders. The electrophotographic powder deposition means (with its photo-receptor and hoppers, e.g.) creates a thin section (image 27) of binder powder with a predetermined shape and dimensions in accordance with a computer aided design (CAD) data of an object and then transfers this binder powder image onto a layer of the primary body-building powder material. The electrophotographic powder deposition means may also produce thin sections of modifier powders with predetermined geometry and material composition distribution (or color pattern) and transfer these modifier powder (toner image) layers onto their corresponding layer of a primary body-building powder material. This transfer of modifier powder images may be conducted before or preferably after the binder powder image is transferred to the same layer of a primary body-building material. The energy means 40 may comprise developer means to "develop" these modifier images (e.g., by setting the colorant-containing resin in a color toner composition) before these colored images are transferred to the surface of a primary body-building layer. If the modifier powders contain other types of additives but no colorant, these powder "images" (thin sections) do not have to be developed and can be transferred to predetermined areas of a primary body-building powder to modify physical properties thereto before, after, or concurrently with the binder powder image transfer step.

[0076] Optional temperature-regulating means (e.g., heaters, coolers, and temperature controllers; well-known in the art, not shown herein) and pump means (not shown) may be used to provide a protective atmosphere and a constant temperature over a zone surrounding the work surface where a part 24 is being built. The heaters may be used to pre-heat the body-building material powder so that when the binder powder is deposited onto a positive region 25 of a layer, the binder powder can be quickly melted and be capable of permeating through the gaps between body-building powder particles in this positive region. The binder fluid provides bridges between these particles and, when the binder is solidified, these particles are bonded and consolidated together. Solidification is accomplished by exposing the binder to an energy means (e.g., heat and/or UV light to cure or harden the binder if the binder is an adhesive) or by exposing the binder fluid to a lower temperature environment below the melting point of the binder. A motion device (not shown) is used to position the work surface 16 with respect to the powder-dispensing device 22, the electro-photographic means (including photo-receptor 18), and the energy means (e.g., light source 40). After a layer of body-building powder, binder and modifier materials is deposited and a cross-section of the 3-D object is built, the powder feeder 22 and the work surface 16 are shifted away from each other by a predetermined distance to get ready for dispensing a next layer of powder. Preferably, it is the work surface that is lowered vertically in the Z-direction so that other devices (including the powder feeder 22, the electro-photographic means, and the energy means will not have to move up in the Z-direction, defined in the Cartesian coordinate system 20 (FIG. 1).

[0077] Electro-photographic Powder Deposition Means

[0078] In one preferred embodiment, the electro-photographic powder deposition means 19, as indicated in FIG. 2, includes a continuous loop photo-receptor belt 58, with means such as motor-powered rollers 30 to drive the belt 58. The belt 58 has a thin layer of photo-conductive or photo-receptive material coated on one side of the belt. The photo-conductive coating is electrically non-conducting unless exposed to a light source. A powder image transferring cycle begins with charging the photo-receptor of the belt by using a charging device 54, of a type known in the art such as a corona charging device. The charged photo-receptor belt is then driven to be positioned before an image projector 66, which creates a latent image 27A of the desired cross-section of the 3-D object (e.g., a cross-section 12A of a coffee cup shown on a CAD computer monitor) by projecting light onto the region to be charged. The image may be formed in a known manner using CRT displays or lasers, as in a laser printer regulated by a computer. The belt 58 is then moved so as to pass by or near a binder powder delivery device 56A. Powder delivery devices are also well-known in the art. A plurality of additional powder delivery devices 56B, 56C, 56D, 56E, etc. may be used to provide different modifier powders (e.g., for cyan, magenta, yellow, and black toner image powders, respectively). A thin layer of preferably charged binder or modifier powder is attracted onto the charged areas (e.g., 27A) of the latent image formed on the belt 58 by image projector 66. Commonly used techniques for transferring the powder to the belt include the use of a magnetic brush device and a triboelectric charging device. This thin layer of binder or modifier powder image is then moved to just above a layer of primary body-building powder material already deposited on a work surface or a previously built layer (e.g., 25A plus 29) supported by this work surface. The primary body-building powder material is preferably pre-charged with charges of a polarity opposite to that in the binder or modifier powder to facilitate binder powder transfer from the belt to the current layer of a primary body-building powder. This work surface 16 sits on a build platform 64 which provides for precise alignment. The platform and the work surface move up and down so that when the binder or modifier powder image is brought into the correct position, the current layer of primary body-building powder material can be brought into a near-contact position with the belt 58 to receive the binder powder therefrom. The image area of the belt 58, after releasing the binder powder, then passes into the belt cleaning device 70, thereby completing one complete electrophotographic powder deposition cycle. Different modifier powders may be formed into separate modifier toner powder images and transferred in a desired sequence to the corresponding layer of a primary body-building powder. Alternatively, different modifier powders may be combined into one composite toner image which is then transferred to the layer of a primary body-building powder.

[0079] The belt 58 is cleaned with each pass by using a cleaner device 70, of a type known in the art which discharges the belt by exposing it to a an intense bright light and which removes any residual particles by brushing or scrapping means. Electrophotographic imaging devices are well-known in the art. Those interested may find useful information in the following U.S. Pat. No.

2,297,691 (Oct. 6, 1942 to C. Carlson), U.S. Pat. No. 3,969,624 (Jul. 13, 1976 to Van Biesen, et al.), U.S. Pat. No. 4,615,606 (Oct. 7, 1986 to Nishikawa) and U.S. Pat. No. 4,652,115 (Mar. 24, 1987 to Palm, et al.).

[0080] Referring to FIG. 5a-5 e, another preferred embodiment of the presently invented method and apparatus includes the operation of a programmable planar powder deposition means (82 in FIG. 5e) which includes an essentially 2-D or plate-like charging device (FIG. 5a or 80A in FIG. 5e) that is capable of providing charges to selected areas of this plate. These areas are programmable and pre-determined by a computer. These areas (the positive region of a layer) are defined by the layer data of a CAD design for the object to be built. The binder or modifier powder is attracted to this positive region only and not to other areas (negative region) of this plate. The bias voltage in each cell can be readily reversed. The charges (e.g., negative charges) produced by a cell are opposite to the charges (e.g., positive charges) provided to the binder or modifier powder when this cell is programmed to attract charges during the formation of a binder powder image. Charges of the same polarity (e.g., both being negative) are produced by this cell by simply reversing the bias charge when it is ready to release the binder powder particles attracted to this cell to a layer of a primary body-building powder material. The bias voltage provided to this plate of a matrix of capacitor cells can be manipulated so that the polarity of charges can be easily reversed once a layer of powder image is released for deposition onto a corresponding layer of a primary body-building powder material.

[0081] As shown in FIG. 5a and 5 b, the plate-like charging device comprises

128 Ausgewählte Patente

basically a dot matrix of capacitors along with their charging circuits. A matrix of minute capacitor "dots" of a substantially uniform size preferably on the level of smaller than 100 μm, further preferably smaller than 10 μm, and most preferably smaller than 1 μm. Each dot can be represented by a cell, schematically shown in FIG. 5(a) and 5(b). An example of a cell circuit diagram, given in FIG. 5(c), comprises two input addresses A and B which send binary bit signals "0" or "1" through an "AND" gate G into a CK terminal of a D-trigger. The output of D is Q, which is connected to transistors TR1 and TR2 for driving a load C (a minute capacitor element). These two transistors alternately provide positive and negative charges to the cell. The gate G, load C, D-trigger, and the transistors TR1 and TR2 together constitute the essential elements of a cell. In a capacitor dot matrix, C is a capacitor that provides charges over a small area, approximately of the cell size. In this circuit, {overscore (O)} is non-Q or opposite to Q with {overscore (O)}="0" when Q="1" and {overscore (O)}="1" when Q="0". Before the start of a powder image formation operation, A and B are in the unselected status (at "0" level), while Q remains at the "0" level (C being "OFF" at the positive charge status) after a "RESET" signal is effected (a short "1" level, then "0"). Logically, the output Q will be "1" (and, hence, C is switched on to provide negative charges) once both the input addresses A and B are "1". The "1" status of the output Q will stay unchanged with C being always in "negative charge" even though either or both of A and B becomes "0". When both A and B of the same cell become "1" again or a new RESET signal comes, the output Q will be changed to "0" again with C providing positive charges. A large number of such cells or capacitor dots can be arranged in a square array as indicated

in FIG. 5(b) by using a micro-electronic fabrication technique such as lithography. As further illustrated in FIG. 5(b), a planar pattern of charged areas in the shape of a capital letter H will be effected when the following pairs of input addresses are in "ON" or "1" status, in the following sequence: (A2,B1), (A2,B2), (A2,B3), (A2,B4), (A2,B5), (A3,B3), (A4,B1), (A4,B2), (A4,B3), (A4,B4), and (A4,B5). When the corresponding cells are switched on, this planar charging device (80A in FIG. 5e) can be brought to a position close to a source of a binder or modifier powder material 84, resulting in attraction of a thin layer of binder or modifier powder particles with positive charges onto the bottom surface of this planar charging plate device 80A, forming a binder or modifier powder "image" of pre-determined shapes and dimensions. In this example, this image of powder particles represents a positive region of an object cross-section designated by the letter H (FIG. 5b). After an H-shaped cross-section is formed, the above cells can be switched off by sending in a new RESET signal or re-selecting the above addresses in that sequence to release this image of binder powder to the corresponding layer of a primary body-building powder material. This implies that the coverage region of this planar image is programmable, in accordance with the CAD-defined cross-section data of a layer.

[0082]FIG. 5(d) shows another example of the logic diagram of cells in a planar charging device that can be conveniently operated. In this diagram, G1, G2, and G3 are the commonly used "NAND" gates in the field of logic circuit design. Herein, G1 is a selectable decoder while G2 and G3 serve as a R-S trigger. In the beginning, all the Cs in the planar charging plate are in the "OFF" status and the RESET terminal remains at the high or "1" level. When both input addresses are selected with "1" level, the functional element C will provide opposite charges and stay in the "ON" status until a new low level RESET signal comes again.

[0083] Referring to FIG. 5e again, the programmable planar powder deposition device 82A comprises a source of positively charged binder or modifier powder 84 inside a chamber 88 which is equipped with a piston-like member 86 that moves the binder or modifier powder up and down to supply a predetermined quantity of binder or modifier powder at a time to the bottom surface of a plate-like charging device 80A. When an image of charges are created at this bottom surface, it attracts a corresponding image of binder or modifier powder to this surface. This plate-like charging device is then moved horizontally to the right along the X-direction and precisely positioned just above a layer 90 of a primary body-building powder material previously deposited by a powder-dispensing device 22A or 22B. At this position, this plate-like device, now designated as 80B, releases the image of binder or modifier powder onto the underlying layer 90 of a body-building material by reversing the cell polarity. The plate-like charging device is then retrieved back to the position designated by 80A and, during this return trip, passes over a cleaning device 70 which removes the residual charges and powder particles on the bottom surface of this plate-like charging device. This device is now ready to prepare another image of binder powder while at the same time energy sources such as a heater and/or UV light 40 are used to consolidate the layer of body-building powder, binder and modifier materials. In the meantime, the work surface 16 is lowered vertically by one

layer thickness distance and the powder-feeder 22B or 22A is activated to move from the right end of the work surface to the left end and back to deposit another thin layer of primary body-building powder material. A new cycle now begins. A multiplicity of powder feeders (e.g.,22A, 22B and more) may be utilized alternately to feed and spread up layers of different primary body-building materials. Further, a multiplicity of binder and modifier powder sources may provide alternate layers of binder and modifier powders to be electrostatically transferred by the planar charging device 80A for forming a multi-material or multi-color object.

[0084]   Powder-Dispensing   Devices (Powder Feeders)

[0085] A wide array of powder-dispensing devices may be used in the present freeform   fabrication   method   and apparatus for feeding the primary body-building material powder. Powder feeders are well-known in the art (e.g., for use in conventional SLS as described in U.S. Pat. No. 4,938,816, Jul. 3, 1990 to Beaman, et al and U.S. Pat. No. 5,316,580, May 31, 1994 to Deckard and for use in 3D powder printing as described in U.S. Pat. No. 5,204,055, Apr. 20, 1993 to Sachs, et al.). We have found it satisfactory to use a device (not shown) to provide a mound of powder with a predetermined volume at a time onto one end of the work surface and move a rotatable drum (22A or 22B in FIG. 2) from this end to another end with a desired spacing between the drum and the work surface. During such a translational motion, the drum also rotates in a direction counter to the translational motion direction, leaving a powder layer thickness being approximately equal to the desired spacing. Preferably, the powder feeder works with a charging device so that the primary body-building powder material dispensed from the feeder 22A or 22B is provided with charges of the polarity opposite to the polarity of the charges in the binder powder image.

[0086] Energy Means

[0087] Several energy means can be used in the practice of the present invention, including utilizing heating sources (infrared, induction heating, dielectric heating, microwave heating, hot-air convective heating, and traditional conduction heating) and/or radiation sources (ultra violet 40, X-ray, Gamma-ray, electron beam, laser beam, ion beam, and plasma). A complete layer of a primary body-building powder material can be pre-heated by selected heat sources disposed near the object-building zone to a temperature (Tpre). For a binder powder that comprises a fusible material composition, this Tpre may be chosen to be above the melting point (Tm) of the fusible material composition so that the binder powder, once deposited onto this layer of body-building powder, is quickly melted to become a binder fluid that permeates through the gaps between powder particles. The heat is then reduced to allow the fluid to be solidified, thereby consolidating or sintering together the powder particles of the primary body-building material.

[0088] For a binder material that is a photo-curable or radiation-curable adhesive, the pre-heat temperature Tpre preferably is not sufficient to significantly initiate a cure reaction, but is sufficient to accelerate the cure reaction once initiated by a photo-initiator (included in the binder powder, e.g.) along with the UV light or other radiation source. Chemical reaction rates are known to increase normally with increasing temperature, but temperature alone may not be sufficient to start out a specific chemical reaction. The pre-heating operation would significantly reduce the light intensity requirement or exposure time that would otherwise be imposed upon the UV light or radiation source. Curing of the binder adhesive in a layer does not necessarily have to be complete before attempting to build a subsequent layer. The cure reaction in a layer may be allowed to continue while other layers are being built, provided the curing is proceeded to an extent that the layer is sufficiently rigid and strong to support its own weight and the weight of subsequent layers.

[0089] Binder Powder, Modifier Powder and Primary Body-Building Powder Materials

[0090] In this method, the photo-curable binder adhesive may consist of such adhesive compositions as a base resin, a hardening or cross-linking agent, a photo-initiator, a photo-sensitizer, and possibly with a reaction accelerator. One or more than one of these compositions (preferably those compositions in a fine solid powder form) may be included as the binder powder to be electro-photographically formed and other remaining compositions as secondary ingredients mixed with the primary body-building powder material to be dispensed one layer at a time by a powder feeder (powder-dispensing means). For instance, the photo-sensitizer (nano-scaled TiO_2 particles) along with other ingredients may be incorporated as the binder powder in the case of photo-curable acrylate materials. These TiO_2 particles, once deposited onto a layer of a mixture of a primary body-building powder material and fused acrylate prepolymer liquid (plus photo-initiators, etc.), may migrate through this layer and help to initiate/accelerate the curing reaction.

[0091] The photo-curable adhesives which can be used in the practice of the present invention are any compositions which undergo solidification under exposure to an actinic radiation. Such compositions comprise usually a photo-sensitive material and a photo-initiator. The word "photo" is used here to denote not only light (preferably UV light), but also any other type of actinic radiation which may transform a liquid adhesive to a solid by exposure to such radiation. A wide variety of photo-curable adhesive resin compositions are available in the art. Examples of this transformation behavior include cationic polymerization, anionic polymerization, step-growth polymerization, free radical polymerization, and combinations thereof. Cationic polymerization is preferable and free radical polymerization is further preferable. One or more monomers may be utilized in the compositions. Monomers may be mono-functional, di-functional, tri-functional or multi-functional acrylates, methacrylates, vinyl, allyl, and the like. The adhesive compositions may comprise other functional and/or photo-sensitive groups such as epoxy, vinyl, isocyanate, urethane, and the like. A large number of examples for photo-curable adhesive compositions can be found in both open literature and patents. For instance, the following U.S. patents provide a good source of these adhesive compositions: U.S. Pat. No. 6,110,987 (Aug. 29, 2000 to Kamata, et al.), U.S. Pat. No. 6,025,112 (Feb. 15, 2000 to Tsuda), and U.S. Pat. No. 5,981,616 (Nov. 9, 1999 to Yamamura, et al.).

[0092] The powder inside a powder feeder 22 may comprise a primary body-building material (fine particles), selected additives (physical or chemical property modifiers that are germane to all layers), and secondary ingredients (selected compositions of an adhesive that are germane to all layers). Those adhesive or modifier ingredients that are required to vary from point to point or layer to layer will be formed into binder or modifier powder images and transferred electrophotographically. In the presently invented method, the primary body-building powder may be composed of one or more than one type of fine particles. These fine powder particles could be of any geometric shape, but preferably spherical. The particle sizes are preferably smaller than 100 μm, further preferably smaller than 10 82 m, and most preferably smaller than 1 μm. The size distribution is preferably uniform. There are three basic types of powders that can be used in the present invention:

[0093] Type A: fine particles of a primary body-building material only. In this type, only primary body-building materials in a fine particle form are included as the ingredients in the powder; no binder or modifier composition being included. All binder or modifier compositions are present in the binder powder image. The primary body-building materials can be selected from polymers, ceramics, glass, metals and alloys, carbon, and combinations thereof. The polymers may be thermoplastic (e.g., polyvinyl chloride) or thermosetting (e.g., epoxy oligomer powder). The binder, including all selected compositions, will be deposited over a complete layer of the primary body-building material and allowed to permeate through the gaps in the powder. The binder in the positive region (corresponding to the desired cross-section) of a layer will be either solidified through cooling (of the binder fluid that contains a fusible material composition) or at least partially cured (for curable adhesive binder chemically cross-linked or otherwise hardened) to bond together the primary body building particles. The powder particles in the negative region will not be exposed to any binder material and will remain as loose or physically separable particles.

[0094] Type B: fine ceramic, metallic, glass, or polymeric particles (as primary body-building materials) each coated with a thin layer of coating comprising selected binder adhesive or modifier compositions. Once a layer of these coated solid particles is deposited, the remaining adhesive compositions of a binder powder image are then deposited, fused (if necessary) and allowed to permeate through the gaps between these particles. These other compositions are then in contact or reacted with the selected compositions in the coating to make a complete binder adhesive. The adhesive in the positive region of a layer is then at least partially cured by the energy means (to bond together body-building particles), leaving the particles in the negative region in a loose or physically/chemically separable state.

[0095] Type C: a mixture of fine particles of primary body-building materials (e.g., a silicon carbide or stainless steel powder) with at least one binder adhesive composition also in a fine powder form (e.g., powdered epoxy oligomer as an adhesive binder resin). The other remaining adhesive compositions (e.g., phot-initiator) are deposited electro-photographically onto a layer of Type C powder mixture and allowed to flow around the fine particles and react with the at least one adhesive composition. The complete adhesive formulation in the positive region of this layer is then at least partially cured to provide inter-particle bonding for those primary body-building particles in the positive region. Again, the powder particles in the negative region

will remain in a separable state.

[0096] The primary body-building material can be selected from a wide variety of materials (polymers, ceramics, glass, metals and alloys, carbons, etc) provided they can be made into a powder form. Most of solid materials can be made into fine particles by using, for instance, a high-energy planetary ball-milling method.

[0097] In each of the above powder types, additional modifier ingredients may be added to impart desired physical and/or chemical properties to the object being built. These ingredients may contain a reinforcement composition selected from the group consisting of short fiber, whisker, and particulate reinforcements such as a spherical particle, ellipsoidal particle, flake, small platelet, small disc, etc. These ingredients may also contain, but not limited to, colorants, anti-oxidants, anti-corrosion agent, sintering agent, plasticizers, etc. Any of these ingredients, when intended to be used in each and every layer of the 3-D object (i.e., germane to all layers), may preferably be included in the primary body-building powder to be dispensed by a traditional powder feeder. Those ingredients that are to be deposited only at selected spots of a layer or selected layers (but not all layers) of an object may be included as a part of a modifier powder. These ingredients will then be electro-photographically formed into a modifier powder image (toner) and transferred to a corresponding cross-section of a primary body-building powder, before or after the binder powder image is transferred. Alternatively, selected ingredients may be combined with a binder powder to form a composite binder-modifier powder image.

[0098] To produce full-color layers, modifier powders may be prescribed to contain colorants. Color toners are well-known in the art. The following U.S. patents provide useful information on color toners and developers: U.S. Pat. No. 5,164,774 (Nov. 17, 1992 to Tomita, et al.), U.S. Pat. No. 5,143,809 (Sep. 1, 1992 to Keneko, et al.), U.S. Pat. No. 5,256,512 (Oct. 26, 1993 to Kobayashi, et al.), U.S. Pat. No. 5,296,325 (Mar. 22, 1994 to Ohtsuka, et al.), U.S. Pat. No. 5,660,959 (Aug. 26, 1997 to Moriyama, et al.), U.S. Pat. No. 5,756,244 (May 26, 1998 to Omatsu, et al.), U.S. Pat. No. 5,721,083 (Feb. 24, 1998 to Masuda, et al.), U.S. Pat. No. 5,919,592 (Jul. 6, 1999 to Yaguchi, et al.), and U.S. Pat. No. 6,004,711 (Dec. 21, 1999 to Bourne, et al.).

[0099] The fact that any material that is available in a powder form can be used in both the traditional selected laser sintering (SLS) and the presently invented full-area sintering technique (FAST) makes both techniques highly versatile. In the present FAST method, additional ingredients may be added by using repeated electrophotographic procedures to impart desired physical and/or chemical properties to the object being built.

[0100] Object-Supporting Work Surface and Motion Devices

[0101] Referring again to FIG. 1, the work surface 16 is located in close, working proximity to the powder-dispensing device 22 and the electrophotographic powder deposition device 19. This work surface 16 has a flat region sufficiently large to accommodate successive layers of the deposited material. The work surface 16 is supported by a build platform 64 which is equipped with mechanical drive means for moving the work surface up and down. The work surface 16 and build platform 64 are preferably contained in a chamber (chamber wall being indicated as 62 in FIG. 2) which is supported by a sturdy base member 72. This member 72 may be optionally equipped with rollers to facilitate moving of the apparatus. The powder-dispensing means 22 is provided with motion devices for moving the powder-dispensing means 22 from one end of the work surface to another end (along the X-direction, e.g.) and for depositing a thin layer of a primary body-building material powder onto the work surface or a previously deposited layer. This can be accomplished, for instance, by allowing the powder-dispensing device to be driven by at least one linear motion device to translate along the X-direction (defined in the X-Y-Z coordinate system 20 of FIG. 2), which is powered by a corresponding stepper motor, and driven to rotate in a direction counter to the translational motion to deposit a layer of powder. The work surface and the electrophotographic powder deposition device can also be moved relative to each other vertically along the Z-direction to make room for the powder-dispensing device 22. Preferably the electrophotographic powder deposition device 19 is driven by a stepper motor to move up and down in the Z-direction relative to the work surface. Motor means are preferably high resolution reversible stepper motors, although other types of drive motors may be used, including linear motors, servomotors, synchronous motors, D.C. motors, and fluid motors. Mechanical drive means including linear motion devices, motors, and gantry type positioning stages are well known in the art. The drive means, motion devices, and planar heat source are preferably subject to automated control by a computer 10, possibly through a hardware control system (14 of FIG. 1)

[0102] These movements will make it

possible for the powder feeder and the electrophotographic powder deposition device to feed successive layers of primary body-building powder, binder powder and modifier powder materials for forming multiple layers of materials of predetermined cross-sections, thicknesses and material compositions, which build up on one another sequentially.

[0103] Sensor means (e.g., optical encoder or laser scanner devices, not shown) may be attached to proper spots of the work surface or the material dispensing devices to monitor the physical dimensions of the physical layers being deposited. Dimensional sensors are well known in the art. The data obtained are fed back periodically to the computer for re-calculating new layer data. This option provides an opportunity to detect and rectify potential layer variations; such errors may otherwise cumulate during the build process, leading to some part inaccuracy. Many prior art dimension sensors may be selected for use in the present apparatus.

Rollendrucksintern

Veröffentlichungsnummer	US8119053 B1
Publikationstyp	Erteilung
Anmeldenummer	US 11/998,151
Veröffentlichungsdatum	21. Febr. 2012
Eingetragen	28. Nov. 2007
Prioritätsdatum	18. März 2004
Erfinder	Bryan Bedal, Ross D. Beers, Steven E. Schell
Ursprünglich Bevollmächtigter	3D Systems, Inc.

Patentzitate (7), Nichtpatentzitate (1), Klassifizierungen (10),Legal Events (1)

Kommentar:

Eine exotische Mischung mehrere Verfahren stellt diese Anmeldung vor. Der Ablauf ist komplex, soll aber laut dem Anmelder 3 D Systems für den Anwender einfacher und die Geräte billiger sein als bisherige Lösungen. Zunächst wird, wie üblich, eine STL oder andere 3D CAD Datei Datei in druckbare Schichten zerlegt. Eine Walze (310) wird dann mit einem sinterfähigen Pulver beschichtet und während des Umlaufes wird dieses Pulver, das aus einer Mischung von einem Metall oder andrem hochschmelzenden Material und einem niedrig schmelzendem Pulver wie z.B. Nylon besteht mit der entsprechenden Schicht durch z.B. eine Halogenlampe belichtet. Dieser Prozessschritt ähnelt somit dem Tonerverfahren in Laserdruckern. Dann wird im nächsten Schritt das entstandene Abbild von der rotierenden, beheizbaren Walze auf eine planare Transportplattform transferiert, wobei überschüssiges. das heißt nicht durch die Hitzewirkung der Lampe ins Nylon eingeschmolzene Material, Sintermaterial entfernt wird. Auch auch hier verläuft das Verfahren analog zum Laserkopieren.

Im nächsten Schritt wird die Plattform, die über einen Scherenhub (242) verfügt mittels Conveyer (244) transportiert,. Dieser „Lift" genannte

Scherenhub hebt die Folie an und verpresst sie mit einer weiteren Plattform, oder bereits gedruckten Lagen, die an dieser Plattform haften. Das 3D Objekt entsteht also hängend von oben nach unten.

Der Hintergrund für den Nutzen der Erfindung sieht die Firma 3D Systems darin, dass die beim Lasersintern notwendige und auch potenziell gefährliche Hochleistungslaserquelle entfallen kann und man stattdessen mit recht konventionellen Hitzequellen arbeiten kann, da man zunächst nur das Trägermaterial (Nylon etc.) aufschmelzen muss.

Das Verfahren wirkt sehr komplex und aufwändig, hat aber seinen Charme. Ob eine so lange Prozesskette in der Praxis tatsächlich reibungslos funktionieren kann muss sich noch beweisen. Die vielen rein mechanischen Prozessschritte muten jedoch etwas antiquiert an.

Die gesamte Beschreibung des Ablaufs ist sehr lang und umfangreich und muss daher vom interessierten Leser aus Platzgründen über die Anmeldenummer beispielsweise unter dem Link http://www.google.com/patents/US8119053 recherchiert werden.

Zeichnungen

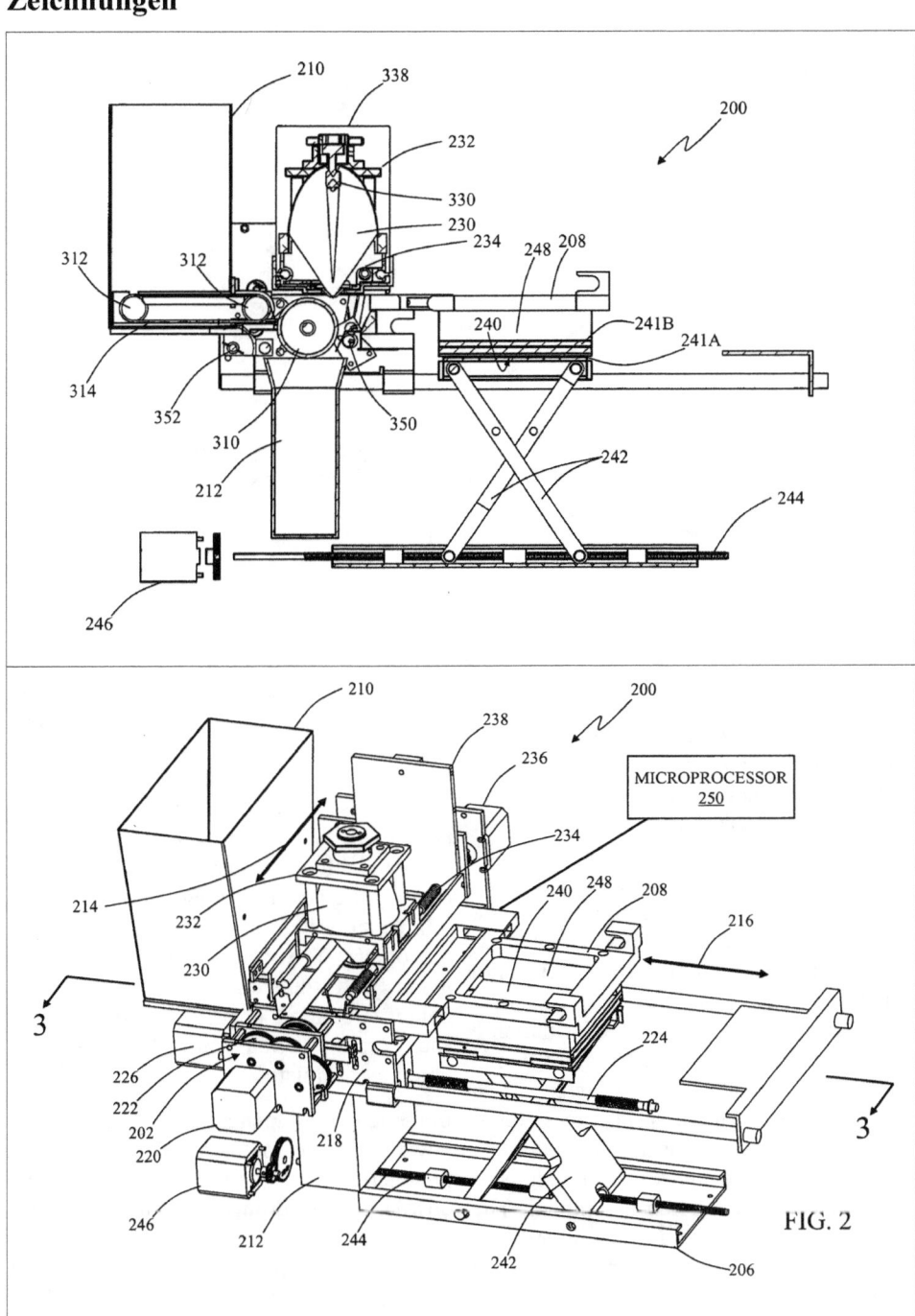

FIG. 2

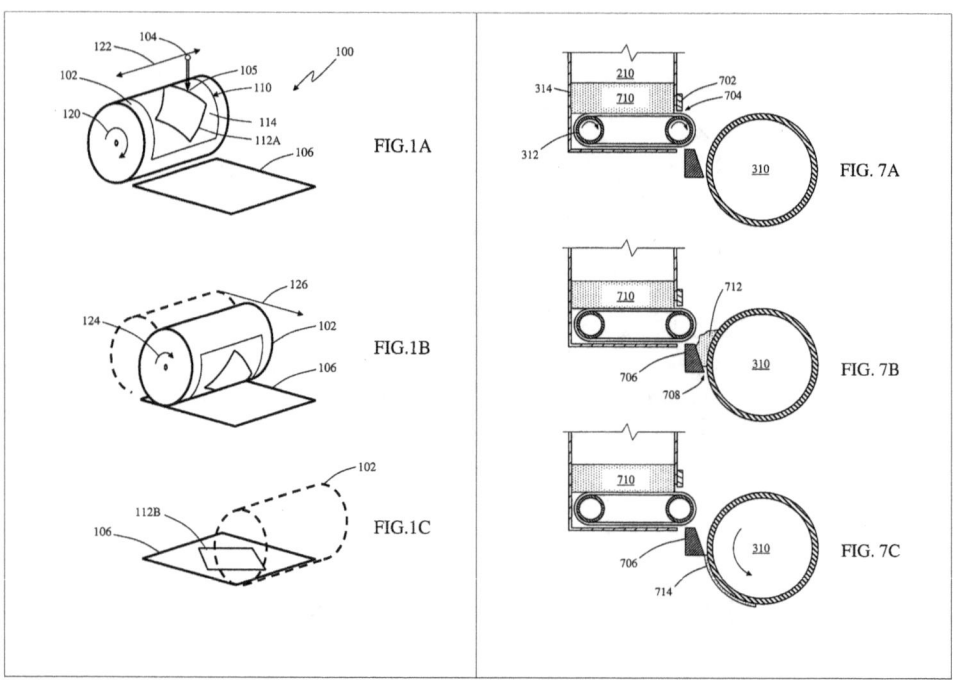

FIG.1A

FIG.1B

FIG.1C

FIG. 7A

FIG. 7B

FIG. 7C

SUMMARY

The invention features a three-dimensional printer (3DP) adapted to construct three dimensional objects from cross sectional layers of the object that are formed on one surface, then subsequently adhered to the stack of previously formed and adhered layers. In the preferred embodiment, the 3DP includes a first surface adapted to receive a bulk layer of sinterable powder; a radiant energy source adapted to fuse a select portion of the layer of sinterable powder to form a sintered image; and a transfer mechanism adapted to concurrently transfer or print the sintered image from the first surface to the object being assembled while fusing the sintered image to the object being assembled. The layer of sinterable powder is preferably a polymer such as nylon that may be fused on a roller or drum, for example, with the energy provided by an incoherent heat source such as a halogen lamp. The transfer mechanism includes one or more actuators and associated controls adapted to simultaneously roll and translate the drum across the object being assembled so as to press and fuse the sintered image to the object. The transfer mechanism may further include a transfixing heater for heating the sintered image and the object immediately before the layer is applied to the object. The process of generating an image and transferring it to the object being assembled is typically repeated for each cross section until the assembled object is completed.

In some embodiments, the 3DP includes a powder applicator adapted to apply a predetermined quantity of sinterable powder to the drum for sintering. In the preferred embodiment, the applicator extracts the sinterable powder from a reservoir and permits the powder to briefly free fall, thereby separating the particles that may have compacted in the reservoir and normalizing the density of the particles applied in layer form to the drum. The powder applicator may further include a blade which, when placed a select distance from and angle relative to the drum, produces a layer of sinterable powder with uniform thickness and density on the drum as the drum is rotated.

In some embodiments, the drum of the 3DP includes a temperature regulator and drum heating element adapted to heat the temperature of the drum at or near the fusing point of the sinterable powder to reduce the energy required by the radiant energy source to print a sintered image from the layer of bulk powder on the drum. The 3DP may further include a first heating element, a second heating element, or both to reduce the energy required to fuse the sintered image to the object being assembled. The first heating element, which is incorporated into a platform assembly on which the object is assembled, for example, is adapted to hold the object at a first predetermined temperature above the ambient temperature. The second heating element is preferably a hot pad adapted to contact and maintain the temperature of the upper surface of the object being assembled at a second determined temperature until the next sintered image is applied to the upper surface. The second determined temperature is less than the melting temperature of the sinterable powder.

The 3DP in some embodiments further includes a layer thickness control processor adapted to regulate the thickness of a sintered image fused to the object being assembled. The layer thickness control processor may vary the thickness of the sintered image before or after transferring to the object being assembled by, for example, varying the quantity of sinterable powder dispensed by the applicator, regulating the position of an applicator blade with respect to the drum, regulating the time and pressure applied by the drum to transfer the sintered image to the object being assembled, compressing the sintered image after it is fused to the object being assembled, and removing excess material from the object being assembled by means of a material removal mechanism.

BRIEF DESCRIPTION OF THE DRAWINGS

The present invention is illustrated by way of example and not limitation in the figures of the accompanying drawings, and in which:

FIGS. 1A-1C are schematic diagrams demonstrating the operation of the three dimensional printer of the first preferred embodiment of the present invention;

FIG. 2 is an isometric view of the three dimensional printer in accordance with the second preferred embodiment of the present invention;

(..)

FIGS. 7A-7C are schematic diagrams demonstrating the operation of the powder applicator in accordance with the second preferred embodiment of the present invention;

Hexapod

Veröffentlichungsnummer	US5401128 A
Publikationstyp	Erteilung
Anmeldenummer	US 07/947,819
Veröffentlichungsdatum	28. März 1995
Eingetragen	18. Sept. 1992
Prioritätsdatum	26. Aug. 1991
Gebührenstatus	Verfallen
Auch veröffentlicht unter	DE69308708D1, 4 weitere »
Erfinder	Paul A. S. Charles, Thomas J. Lindem
Ursprünglich Bevollmächtigter	Ingersoll Milling Machine Company

Patentzitate (25), Nichtpatentzitate (8), Referenziert von (69),Klassifizierungen (15), Legal Events (11)

Externe Links: USPTO, USPTO-Zuordnung, Espacenet

Kommentar: Eine verfallene Patentanmeldung aus 1991für eine Werkzeugmaschine auf Basis eines Hexapoden. Evtl. kann man mit den verfallenen Ansprüchen etwas anfangen. Hier kann niemand mehr Ansprüche stellen. Ingersoll Milling Machine ging 2002 in die Insolvenz. Evtl. sind die Patentgebühren daraufhin nicht mehr bezahlt worden. Interessant ist auch die Verankerung der Arme/Beine oben am Gehäuse. Es ermöglicht aber eine bessere Zugänglichkeit zum Werkstück und weniger Totraum als die derzeitige Konstruktion nach dem Rostock Prinzip. Die Arme sind mit Hubkolben statt über umgelenkte Riemenantrieb ausgestattet. Der Titel der Publikation lautet übersetzt ungefähr „Achteckiger Hexypod mit dreieckiger Servo Strebenaufhängung"

U.S. Patent   Mar. 28, 1995   Sheet 1 of 5   5,401,128

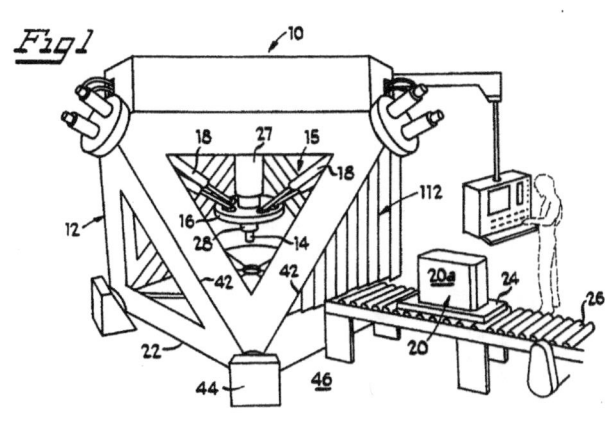

Fig 1

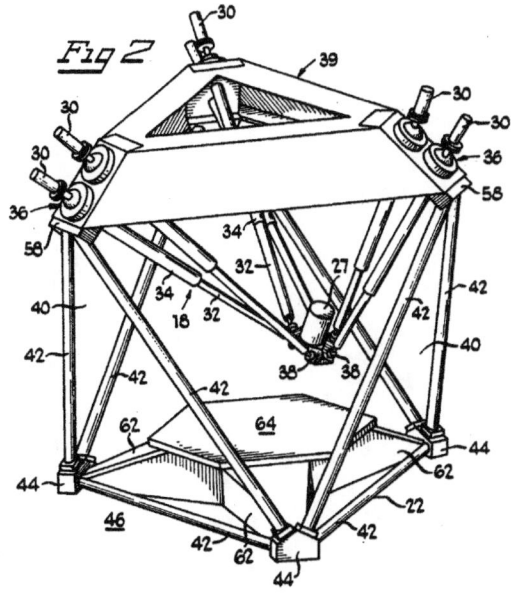

Fig 2

3D Scanner

Einrichtung zum Scannen dreidimensionaler Objekte

Veröffentlichungsnummer	US8294958 B2
Publikationstyp	Erteilung
Anmeldenummer	US 12/299,349
PCT-Nummer	PCT/GB2007/001610
Veröffentlichungsdatum	23. Okt. 2012
Eingetragen	3. Mai 2007
Prioritätsdatum	4. Mai 2006
Auch veröffentlicht unter	EP2024707A1, EP2024707B1,US20090080036, WO2007129047A1
Erfinder	James Paterson, Ronald William Daniel,David Claus, Andrew Fitzgibbon
Ursprünglich Bevollmächtigter	Isis Innovation Limited

Patentzitate (34), Nichtpatentzitate (19), Referenziert von (1),Klassifizierungen (15), Legal Events (1)

Externe Links: USPTO, USPTO-Zuordnung, Espacenet

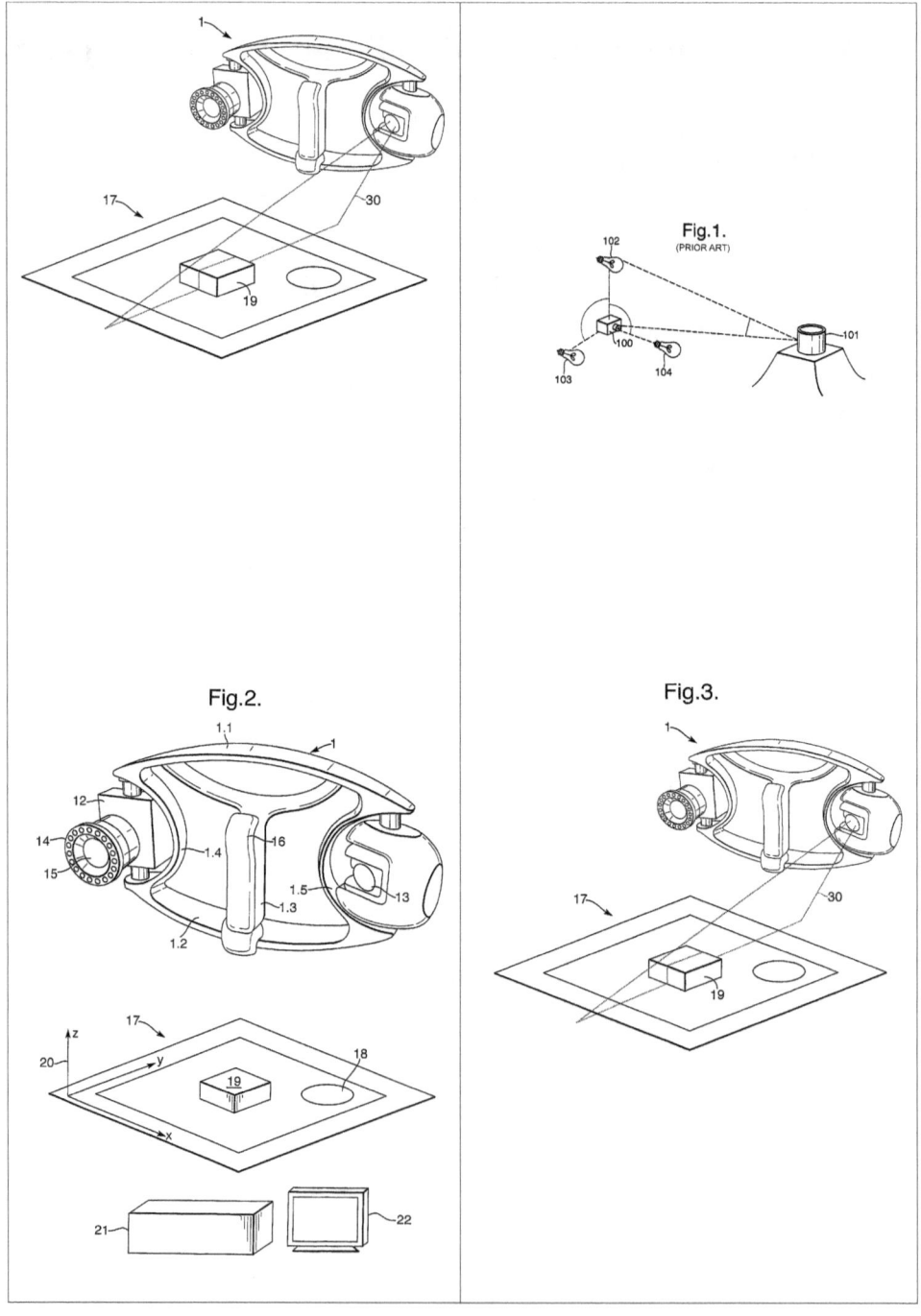

Fig.1.
(PRIOR ART)

Fig.2.

Fig.3.

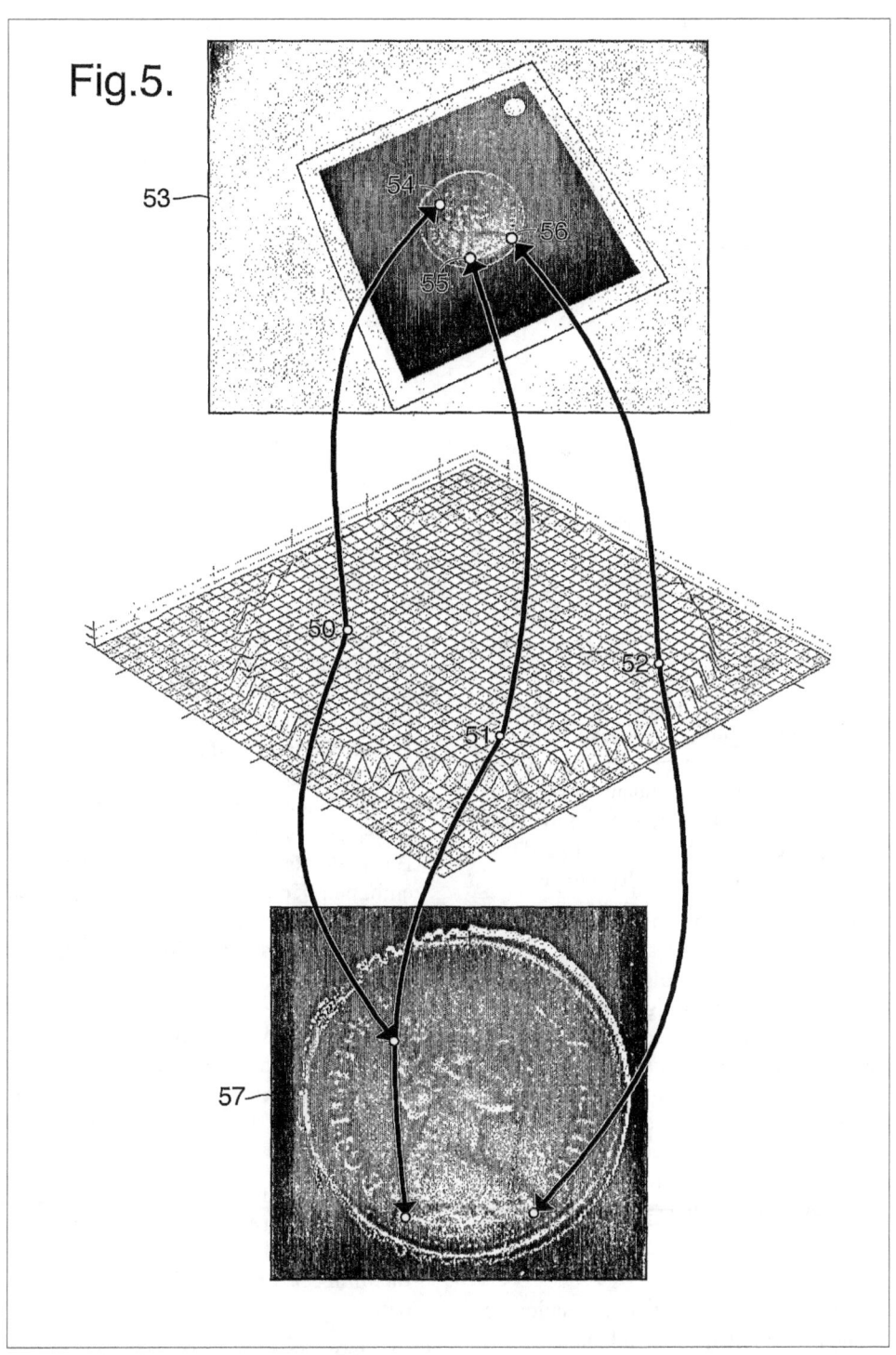

Fig.5.

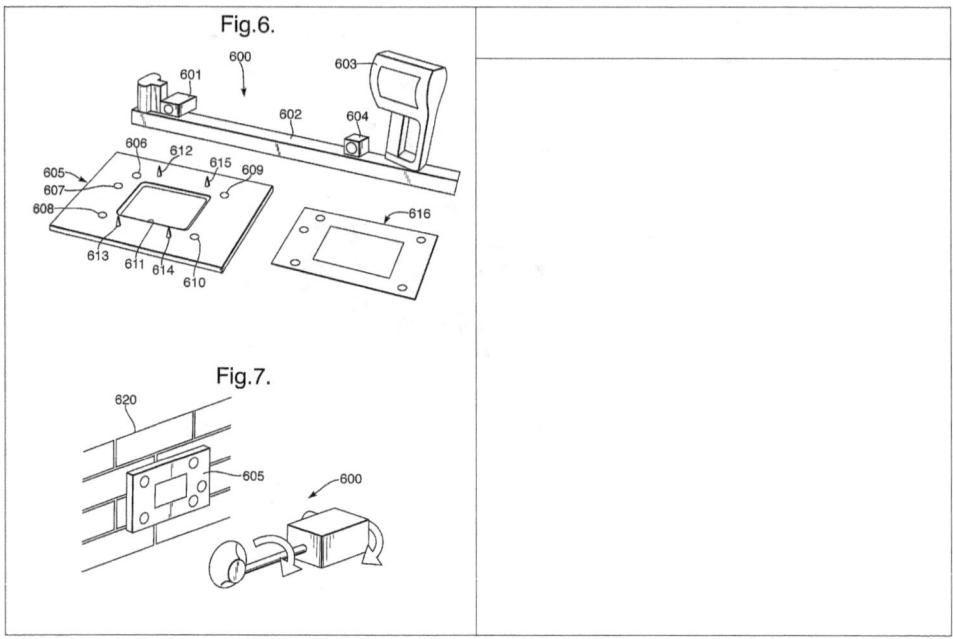

Fig.6.

Fig.7.

Beschreibung

[0001] Die Erfindung betrifft Scaneinrichtungen und insbesondere ein Verfahren und eine Scaneinrichtung, die geeignet zum Scannen von dreidimensionalen Objekten und zum Ermöglichen der Rekonstruktion eines 3D-Models eines realen Objektes ist.

[0002] Die hier als Hintergrund beschriebenen Ansätze könnten ausgeführt oder verfolgt werden, sind jedoch nicht unbedingt Ansätze, die bereits verfolgt oder in Betracht gezogen wurden. Wenn also nicht abweichend angegeben, sind diese als Hintergrund beschriebenen Ansätze nicht Stand der Technik für die Ansprüche dieser Anmeldung und es wird auch nicht eingeräumt, dass diese aufgrund ihrer Aufnahme zum Stand der Technik zählen.

[0003] Dreidimensionale Scaneinrichtungen beinhalten im Üblichen einen zweistufigen Prozess der Datenerfassung, gefolgt von der Rekonstruktion eines 3D-Bildes. Die Datenerfassung umfasst eine Sammlung von Rohdaten, die zur Erzeugung der anfänglichen Geometrie des gescannten Objektes verwendet wird. Übliche Beispiele der Datenerfassungsvorgänge enthalten Kontaktverfahren (z. B. eine mechanische Sonde, welche die Oberfläche eines Objekts abtastet) und Nicht-Kontaktverfahren (wie beispielsweise Abbildungs- und Abstandsmessverfahren). Bei Rekonstruktion werden diese Daten in geeignete Daten für 3D-computer-aided design (CAD) und/oder Animationsanwendungen verarbeitet.

[0004] Eine dieser Nicht-Kontaktverfahren umfasst eine optische Technik, bei der ein strukturiertes Lichtmuster auf ein Objekt projiziert wird.

"A low cost 3D scanner based an

structured light", von C. Rocchini et al. in Eurographics 2001, Ausgabe 20, Nr. 3 beschreibt ein Verfahren auf Grundlage von strukturiertem Licht mit einer optischen Licht-Kontaktscantechnik. Ein strukturiertes Lichtmuster wird auf ein Objekt projiziert und ein Sensor nimmt ein Bild des gestörten Lichtmusters aus der Oberfläche des Objekts auf. Tiefeninformationen werden dann durch Triangulation zwischen dem Sensor, dem Emitter und dem gesampelten Punkt rekonstruiert. Solch ein Verfahren erfordert einen hohen Grad an Genauigkeit insbesondere zwischen den überwachten Positionen des Emitters, des Sensors und einem gesampelten Punkt, so dass die Tiefeninformationen genau rekonstruierbar sind. Solche Verfahren erfordern hoch entwickelte Einrichtungen und bei jeder Verwendung eine Kalibrierung.

[0005] "Calibration-Free Approach to 3D Reconstruction using Light Stripe Projections an a Cube Frame" von Chu et al., Proceedings of the

Third International Conference an 3D Digital Imaging and Modelling 2001 beschreibt ein Beispiel eines Systems, bei dem ein Lichtebenenprojektor Licht auf ein Objekt innerhalb eines dreidimensionalen Rahmens projiziert. Der dreidimensionale Rahmen enthält eine Mehrzahl von lichtemittierenden Dioden (LED), die an den Ecken des Rahmens angeordnet sind. Ein Benutzer führt einen Laser-Streifenbilder über das Objekt. Ein Sensor in Form einer Kamera nimmt ein Bild des gestörten projizierten Lichtmusters auf, wie es von dem Rahmen und der Oberfläche des Objektes innerhalb des Rahmens reflektiert wird. Das aufgenommene Bild enthält auch Bilder der LED's, die dem Sensor die Detektion des

Würfelrahmens ermöglichen.

[0006] Die Kamera, wie sie in der Implementierung verwendet wird, die von Chu et al. beschrieben wird, erfordert keine intrinsische Kalibrierung.

Es ist jedoch darauf zu achten, dass ein genauer Würfelrahmen gebaut wird und eine komplexe Bildverarbeitung ist erforderlich. Außerdem kann der Würfelrahmen das Objekt blockieren. Insbesondere können Schatten an den Kanten des Würfels auftreten, welche die Präzision, mit der Kanten detektiert werden können auf lediglich ein paar Pixel verringern kann. Daher ist das Bild des Würfels wahrscheinlich nicht würfelförmig und robuste Statistiken müssen zur Auffindung der Bildmerkmale verwendet werden.

[0007] Die US 6493095 beschreibt einen optischen 3D-Digitalisierer zum Digitalisieren eines Objekts, welches eine Weißlichtquelle aufweist, die weißes Licht abgibt, eine Projektionslinse, die das Weißlicht auf das Objekt projiziert, wobei das Objekt eine vollständig beleuchtete Seite aufweist,

eine Beugungseinrichtung die ein Beugungsmuster in dem durch die Projektionslinse projizierten Lichtes bildet und erste und zweite Kameras, die seitlich von der Projektionslinse positioniert sind und bezüglich einander in Winkeln ausgerichtet sind, so dass die Kameras komplementäre Sichtfelder aufweisen, die auf die beleuchtete Seite des Objektes gerichtet sind und einander über cinc Messungstiefe des Objektes überlappen. Die KamerasDieser Text wurde durch das DPMA aus Originalquellen übernommen. Er enthält keine Zeichnungen. Die Darstellung von Tabellen und Formeln kann unbefriedigend sein.

haben jeweils Videoausgänge zum Bereitstellen eines Videosignals, welche die komplementären Bilder des Objektes mit einem gemeinsamen Bildanteil als Ergebnis der teilweise überlappenden Sichtfelder darstellen.

[0008] Die US 5307153 betrifft eine dreidimensionale Messeinrichtung mit einem Mehrfach-Schlitzprojektor der eine Betätigungseinrichtung zum Ersetzen wenigstens eines von ersten und zweiten Beugungsgittern in lediglich einer geringen Distanz in Richtung senkrecht zu den Schlitzlichtern aufweist. Eine Bilderfassungseinrichtung enthält eine Bild-Arithmetikeinheit zum Wechseln, immer dann wenn ein kodiertes Multi-Schlitzlichtmuster verändert wird, eine Wichtung der binärisierten Bildsignale und zum Addieren der letzten gewichteten binärisierten Bildsignale der Bildsignale des letzten addierten Ergebnisses, welches aus einem Bildspeicher gelesen wurde mit neuen gewichteten binärisierten Bilddaten.

[0009] Die US 5848188 beschreibt eine hochauflösende Formmessungseinrichtung mit einer Scaneinrichtung, die ein Scannen eines zu vermessenden Objektes bei einer hohen konstanten Rotationsgeschwindigkeit erlaubt. Die Formmessungseinrichtung enthält eine Laserlichtquelle zum Erzeugen eines blitzenden Laserlichtes, einen Musterspeicher zum Speichern von verschiedenen Blitzmustern des Laserlichts, einen Polygonspiegel als Scaneinrichtung und eine CCD-Kamera zum Aufnehmen eines Bildes, welches auf dem Objekt durch das Scannen des Laserlichts gebildet wird, wobei die Form des Objekts gemäß den Bilddaten von der CCD-Kamera errechnet wird.

[0010] Die Formmessungseinrichtung enthält ferner einen Motor zum Drehen des Polygonspiegels und einen Fotodetektor zum Erzeugen eines PD-Signals wenn die Phase der Rotation des Polygonspiegels eine bestimmte Phase einnimmt. Der Polygonspiegel wird derart rotiert, dass er das Objekt mehrmals in der Aufnahmezeit pro Bild der CCD-Kamera scannt, in dem eines der Blitzmuster verwendet wird, und das Blitzmuster wird durch das PD-Signal zurückgesetzt und wiederholt während der Aufnahmezeit pro Bild verwendet. Ferner kann die Anzahl von Scans für ein tatsächliches Richten des Laserlichts auf das Objekt aus einer Anzahl von mehreren Scanzahlen gewählt werden, die in der Aufnahmezeit pro Bild erfolgt, um die Laserlichtquantität einzustellen.

[0011] Auf Grundlage des Voranstehenden gibt es einen klaren Bedarf für eine Scaneinrichtung, welche den erforderlichen Grad der Genauigkeit erreicht, ohne komplexe Verarbeitungsschritte oder komplexe Referenzrahmen zu erfordern.

[0012] Die vorliegende Erfindung wird ein beispielhafter Weise und nicht in beschränkender Weise in den Figuren und den beiliegenden Zeichnungen dargestellt, wobei: [0013] Fig. 1 ein schematisches Diagramm einer Ausführung eines Scansystems ist; und [0014] Fig. 2 ein Ablaufdiagramm ist, welches den Betrieb einer Ausführungsform der Scaneinrichtung zeigt.

[0015] Ein Verfahren und eine Einrichtung zum Scannen von 3D-Objekten werden beschrieben. In der nachfolgenden Beschreibung werden verschiedene bestimmte Details zum

Zweck der Erklärung dargestellt, um ein tiefgehendes Verständnis der vorliegenden Erfindung zu vermitteln. Es ist dem Fachmann jedoch klar, dass die Erfindung auch ohne diese Details ausführbar ist. In anderen Beziehungen werden wohlbekannte Strukturen und Einrichtungen in Blockdiagramm-Darstellung gezeigt, um eine unnötige Verkomplizierung der vorliegenden Erfindung zu verhindern.

[0016] Die oben beschriebenen Bedürfnisse und andere Bedürfnisse und Aufgaben, die aus der nachfolgenden Beschreibung klar werden, werden durch ein Scansystem zum Erstellen eines dreidimensionalen Modells eines Objektes erfüllt, welches gemäß einem Aspekt einer Scaneinrichtung mit einem Emitter zum Projizieren von Licht und einem Sensor zum Aufnehmen von Bildern aufweist, wobei der Emitter und der Sensor während der Benutzung relativ zueinander in einer festen Position stehen und einer Scanvorlage (scanning template) mit einer bekannten zweidimensionalen Vorlage. Verarbeitungsmittel erzeugen Daten zur Ermöglichung der Konstruktion eines dreidimensionalen Modells eines Objektes, welches zwischen der Scaneinrichtung und der Scanvorlage platziert wird. Die Verarbeitungsmittel erzeugen im Gebrauch Informationen über das Objekt relativ zu der Scanvorlage, wobei die Informationen aus dem gleichen Bild erzeugt werden, auf welches das projizierte Licht durch den Emitter projiziert wird.

[0017] Fig. 1 zeigt eine Ausführungsform des Scansystems. Das Scansystem umfasst eine Scaneinrichtung 2 , ein Scanvorlagenmittel 4 und eine Scansoftware, die in der Scaneinrichtung 2 oder in einer separaten Verarbeitungseinheit 6 gespeichert sein kann, mit der die Scaneinrichtung 2 verbunden ist. Die Verabeitungseinrichtung 6 ist üblicherweise ein Personalcomputer PC. Die Software enthält Details der Scanvorlage 4 .

[0018] Die Scaneinrichtung 2 weist einen Emitter 20 und einen Sensor 22 auf. Die Positionsbeziehung zwischen dem Emitter 20 und dem Sensor 22 ist gemäß einer groben Abschätzung bekannt, die einen Eingabe "seed" für den Berechnungsalgorithmus bereitstellt. Diese Positionsbeziehung ist während eines Scans feststehend. Wenn nicht im Gebrauch, können Emitter 20 und Sensor 22 beispielsweise weggeräumt werden und im Gebrauch relativ zueinander zu einer bekannten Relativposition gebracht werden. Zusätzlich oder alternativ können Emitter und Sensor zueinander zwischen mehreren feststehenden Positionen bewegt werden. In der Ausführungsform in Fig. 1 ist der Emitter 20 gegenüber dem Sensor 22 durch einen Arm 24 fixiert. Der Emitter 20 und der Sensor 22 haben bezüglich einander feststehende Positionen, die durch den Arm 24 vorgebbar und kontrollierbar sind.

[0019] Der Emitter ist üblicherweise ein Laserpointer und kann Mittel aufweisen, um das von dem Laser projizierte Licht zu strukturieren. Für beste Ergebnisse muss das durch den Emitter 20 projizierte Licht dünn und gradlinig sein und eine Ebene von Licht bilden, die im Gebrauch auf das zu scannende Objekt Seite 4 --- ()

fällt. Der Sensor ist eine Bildaufnahmeeinrichtung, beispielsweise eine CCD-Einrichtung wie eine Digitalkamera. Üblicherweise ist die Größe der Scaneinrichtung derart bemessen, dass sie für Hand-Held-Anwendungen geeignet ist. Alternativ kann die Scaneinrichtung 2 fest montiert sein, beispielsweise auf einem Dreibeinstativ oder ähnlichem.

[0020] Das Scanvorlagenmittel 4 weist eine zweidimensionale planare Oberfläche 40 auf, die darauf eine zweidimensionale Vorlage 22 bekannter Form und Größe aufgedruckt hat. Wie in Fig. 1 gezeigt ist das zu scannende Objekt 8 auf der Scanvorlage 42 des ZweiDimensionalenscanvorlagenbauteils 4 derart platziert, dass das Objekt 8 zwischen der Scaneinrichtung 2 und der Scanvorlage 42 platziert ist. In Fig. 1 ist die Scanvorlage 42 als ein auf eine Ebene planare rechteckige Oberfläche 40 gedrucktes Rechteck gezeigt, wie z. B. auf ein Stück Papier.

Andere Formen von Scanvorlagen können verwendet werden.

[0021] Das Scanvorlagenteil kann von einem Benutzer hergestellt werden, z. B. in dem ein Benutzer den PC 6 anweist, eine Scanvorlage 42 auf eine ebene Oberfläche, wie z. B. ein Stück Papier oder eine Karte zu drucken. Der Benutzer kann auch in der Lage dazu sein, die Farbe der Scanvorlage 42 derart auszuwählen, dass sie mit dem zu scannenden Objekt einen Kontrast bildet. In jedem Fall werden die Details der erzeugten oder bereit gestellten Scanvorlage für den Zugriff für die weitere Verarbeitung gespeichert.

[0022] In dem beschriebenen Scansystem wird die absolute Position der Scaneinrichtung 2 bezüglich des Objektes 8 nicht benötigt, da der Sensor 22 Informationen über das Objekt 8 bezüglich der Ebene der Scanvorlage 42 erzeugt. Diese Information wird aus dem gleichen Bild erzeugt, auf den

projizierte Lichtmuster durch den Emitter 20 projiziert wird, welche bezüglich dem Sensor 22 in einer fixierten Position ist. Die Position des Lichtstreifens auf das Objekt 8 ist auch bezüglich des Bildes des Streifens und bezüglich der Ebene der Scanvorlage 42 bekannt.

[0023] Das durch den Sensor 22 aufgenommene Bild enthält ein Bild der Scanvorlage 42 und stellt eine gute Abschätzung der Position des Sensors 22 bezüglich des Objekts 8 bereit. Das Scansystem ist ebenfalls mit Details des strukturierten Lichtmusters programmiert, welches durch den Emitter 20 zu emittieren ist. Das Bild 12 des Referenzstreifens von dem Emitter 20 wird auf der Oberfläche der Scanvorlagenebene 42 und auf dem Objekt 8 bereitgestellt. Daher ist die Position des Referenzstreifens 12 bezüglich der Ebene der Scanvorlage 42 äußerst akkurat bekannt und deutlich genauer als die Genauigkeit der Kalibrierung des Sensors 22 .[0024] Dort wo der Referenzstreifen 12 die Scanvorlagenebene 42 schneidet wird dieser Bereich als Disparitätszone 14 bezeichnet und die Position des Referenzstreifens 12 in der Disparitätszone 14 gibt uns die Position der Oberfläche des Objekts.

[0025] Solch eine Ausführungsform ersetzt die absolute Genauigkeit bei einem System in dem alle Fehler erster Ordnung ausgeräumt sind (sogar solche in dem Sensor) da alle Messungen bezüglich derselben planaren Referenz vorgenommen werden. [0026] Zur Verwendung des Scanners erfordert die Kamera zunächst eine Kalibrierung. Diese kann erreicht werden, indem ein Algorithmus der kleinsten Quadrate verwendet wird und vorzugsweise ein rekursiver Algorithmus der kleinsten

Quadrate. Die Kamera wird kalibriert in dem der Benutzer eine Anzahl von Bildern eines Kalibrierungsmusters aufnimmt, die durch den Drucker des Benutzers ausgedruckt werden. Verfahren zur Kalibrierung von Kameras auf Grundlage solcher ebenen Bilder sind wohl bekannt und treten in der frei zugänglichen Literatur auf, beispielsweise in J.Y. Bouguet's "Camera Calibration Toolbox for Matlab" gefunden auf http:www.vision.Caltech.edu/bouguetj/calib_doc/index.html.

[0027] In einer bevorzugten Ausführungsform der Erfindung gibt der Benutzer dem System eine ungefähre Größe des zu scannenden Objektes vor, dieser Schritt ist jedoch für den richtigen erzielten Effekt nicht unbedingt erforderlich. Zusätzlich gibt es verschiedene Abwandlungen welche einen bequemen Mechanismus zum Automatisieren der Eingabe einer Größenabschätzung ermöglichen, wie die Bereitstellung von verschiedenen Scanvorlagen von verschiedener Größe und Form, wobei der Benutzer diejenige auswählt, die am besten zu dem zu scannenden Objekt passt. In dem ein einzigartiges Identifizierungsmuster auf dem Rand jeder verschiedenen Scanvorlage angeordnet wird, können der Scanner und der Prozessor konfiguriert sein, um die Größe der Vorlage zu identifizieren, die durch den Benutzer ausgewählt wurde, gleichzeitig mit dem Scannen des Testobjekts.

Dies stellt automatisch die Eingabe der geschätzten Objektgröße bereit. Verschiedene Abwandlungen dieses Ansatzes bestehen und können von jemandem, der mit Gerätesoftwaredesign (instruments of the design) vertraut ist, einfach aufgefunden werden.

[0028] Weitere Details bezüglich der Funktionsweise des Scansystems werden nun mit Bezug auf Fig. 2 beschrieben. Der Emitter 20 projiziert Licht 10 auf das Objekt 8 und das Scanvorlagenbauteil 4 . Der Sensor 22 empfängt ein aufgenommenes Bild (201), welches das Bild des Referenzstreifens 12 enthält, der auf die Oberfläche 40 des Scanvorlagenbauteils 4 projiziert ist und, innerhalb der Disparitätszone 14 , die Schnittstelle des Lichts mit dem Objekt 8 . In Antwort darauf findet (202) der Prozessor, der mit den Details der Form der Scanvorlage programmiert ist einen Teil der Scanvorlage, z. B. findet der Prozessor die Ecken der Scanvorlage 42 . Der Prozessor verwendet dann einen Konvolutionsoperator zum Auffinden der Seiten der Scanvorlage 42 .

Um dies zu erreichen, traversiert (203) das System die Seiten der Scanvorlage 42 unter Verwendung eines Vorwärts-Rückwärts adaptiven erweiterten Kalmanfilters.

[0029] Sobald der Prozessor die Seiten der Scanvorlage 42 aufgefunden hat, ist die Position der Scanvorlage relativ zu der Scaneinrichtung dem Prozessor bekannt, da der Prozessor Hinweise bezüglich der Form und der Dimensionen der Scanvorlage hat. Das System bestimmt (204) daher die Position der Scanvorlage bezüglich der Kamera und erzeugt Statistiken der Position. Der Sensor braucht nicht senkrecht zu der Scanvorlagenebene zu sein um zu wirken, da die gesamte Geometrie, auf der die Erfindung basiert, auf schrägen Linien basiert.

[0030] Das System erzeugt dann (205) vorrangige Verteilung des Schnitts des Referenzstreifens 12 mit der Ebene der Scanvorlage 42 unter Verwendung eines

Näherungsmodells des Lichtmusters und der Statistiken die vorhergehend erzeugt wurden. Der Prozessor findet den projizierten Referenzstreifen 12 in dem Bild durch Verarbeiten des Bildes entlang der Seite der Scanvorlage. Sagen wir beispielsweise, dass der Prozessor von der unteren linken Ecke der Scanvorlage 42 , wie in Fig. 1 gezeigt, ausgeht und entlang der Seite A der Scanvorlage voran schreitet. Ein Kantendetektionsalgorithmus kann verwendet werden, um den Punkt B des Referenzstreifens 12 zu identifizieren, welcher die Seite A der Scanvorlage

schneidet. Sobald dieser Punkt B auf jeder Seite der Scanvorlage 42 aufgefunden wird, sucht der Prozessor entlang der Linie B–B innerhalb der Disparitätszone 14 und sucht nach Diskontinuitäten in dem Referenzstreifen 12 . Der Prozessor schätzt dann die Dicke des Objektes 8 ab, welches durch Näherung der Linie abgebildet wird.

[0031] Die vorrangige Verteilung aus Schritt 205 wird verwendet, um einen Bayesfilter (206) anzutreiben um eine affine Näherung des Lichtmusters auf der Scanebene außerhalb der Scanvorlage 42 zu rekonstruieren. Die affine Näherung des Scanmusters wird dann verwendet (207) um den Schnitt der Scanvorlage 42 mit dem zu scannenden 3D-Objekt vorauszusagen.

[0032] Vorherige 2D-Daten werden dann projektiv (208) für das erwartete Bild erzeugt, unter Verwendung von Schritt 207 auf Basis irgendeines vorherigen Scans oder des ersten Scans, wobei die Begrenzungen an den erwarteten Bildpositionen angeordnet sind. Die vorhergehenden Daten für die Schnittstelle des strukturierten Lichtmusters mit dem zu scannenden

Objekt werden dann als Vorlage für einen adaptiven 2D erweiteren KalmanFilter verwendet (209), um das strukturierte Lichtmuster auf dem zu scannenden Objekt zu detektieren.

[0033] Die Statistiken der nachgeordneten Daten, die durch den Kalmanfilter in Schritt 208 erzeugt werden, werden dann verwendet (210),

um vorläufige Daten der 3D-Geometrie des zu scannenden Objektes zu erzeugen. Die vorhergehenden Daten der Objektgeometrie werden dann geupdated (211) unter Verwendung einer Bayes-Updateregel zur Erzeugung einer nachgeordneten Version eines 3D-Modells des gescannten Objekts, und man ist bereit für die nächste Bildaufnahme.

[0034] Diese Schritte werden dann für einen Bereich von projizierten Lichtlinien wiederholt (212) wobei inkrementale Veränderungen beim Auftreffen auf die Scanvorlage stattfinden, bis das gesamte Objekt auf der Scanvorlage 42 gescannt wurde. Die erzeugten Daten können dann zur Bereitstellung an eine Modellierungsanwendung verarbeitet werden, wie z. B. eine herkömmliche 3D-CAD oder Animationsanwendung oder ähnliches. [0035] In dieser Beschreibung wird "vorläufig" (prior) als Kurzform für "vorläufige Verteilung" verwendet und nachgeordnet (posterior) ist eine Kurzform für eine "nachgeordnete Verteilung". Die vorläufige ist eine Wahrscheinlichkeitsbeschreibung des Wissenszustandes bevor eine Messung getätigt wird und nachgeordnet ist der Wissenszustand nach der Messung. Im Allgemeinen kommt man von der Vorgeordneten zu der Nachgeordneten unter Verwendung von Bayes-Regeln.

[0036] So wird eine kalibrierte Scaneinrichtung bereitgestellt, die offline unter Verwendung eines Algorithmus kalibriert wird, der bekannte Parameter und Charakteristika des Systems verwendet (sogenannte "intrinsische Parameter"). Die "extrinsischen Parameter" des Sensors, d. h.solche Variablen, die nicht intrinsisch innerhalb des Designs des Sensors festgelegt sind (z. B. die Position des Objektes) werden bei der Verwendung unter Gebrauch einer Scanvorlage bekannter Größe und Form geschätzt, die von einem Benutzer hergestellt sein kann.

[0037] Der beschriebene Scanner kann für viele Zwecke verwendet werden, z. B. für Heimgebrauch (z. B. Scannen eines geliebten Gegenstandes, wie z. B. eines Ziergegenstandes oder eines Juwelenstückes), in dem Spiel- und Unterhaltungsbereiche (z. B. zum Scannen und Rekonstruieren einer Figur zur nachfolgenden Animation), zum akademischen Gebrauch (z. B. zum Scannen von Antiquitäten) oder für den medizinischen Gebrauch.

[0038] Die hier gegebenen Beispiele sind nur beispielhaft und nicht in beschränkender Weise zu verstehen.

Weitere relevante Patente

Hier in Kurzform Patente, die häufig zitiert werden und daher für 3D Drucker relevante Informationen enthalten:

Patent	Einge-tragen	Antrag-steller	Titel
US3016451	04.06.57	The Auto Arc-Weld Mfg. Co.	Electrode feed roll means
US3841000	26.03.73	W Us Simon	Reel closure
US3917090	02.11.73	Pitney-Bowes, Inc.	Postage meter tape recepticle system
US4152367	25.07.78	Bayer Aktiengesells chaft	Branched polyaryl-sulphone/polycarbonate mixtures and their use for the production of extruded films
US4575330	08.08.84	Uvp, Inc.	Apparatus for production of three-dimensional objects by stereolithography
US4665492	02.07.84	Masters; William E.	Computer automated manufacturing process and system
US4749347	21.10.86	Valavaara; Viljo	Topology fabrication apparatus
US4844373	18.12.87	Fike, Sr.; Richard A.	Line storage and dispensing device
US4898314	20.10.88	International Business Machines Corporation	Method and apparatus for stitcher wire loading
US4928897	18.01.89	Fuji Photo Film Co., Ltd.	Feeder for feeding photosensitive material
US4961154	02.06.87	Scitex Corporation Ltd.	Three dimensional modelling apparatus
US5031120	22.12.88	Barequet; Gill	Three dimensional modelling apparatus

US5059266	23.05.90	Brother Kogyo Kabushiki Kaisha	Apparatus and method for forming three-dimensional article
US5121329	30.10.89	Stratasys, Inc.	Apparatus and method for creating three-dimensional objects
US5134569	26.06.89	Masters; William E.	System and method for computer automated manufacturing using fluent material
US5136515	07.11.89	Helinski; Richard	Method and means for constructing three-dimensional articles by particle deposition
US5140937	23.05.90	Brother Kogyo Kabushiki Kaisha	Apparatus for forming three-dimensional article
US5149548	18.06.90	Brother Kogyo Kabushiki Kaisha	Apparatus for forming three-dimension article
US5169081	06.01.92	Draftex Industries Limited	Strip handling apparatus and method
US5204055	08.12.89	Massachusetts Institute Of Technology	Three-dimensional printing techniques
US5216616	01.12.89	Masters; William E.	System and method for computer automated manufacture with reduced object shape distortion
US5257657	08.07.92	Incre, Inc.	Method for producing a free-form solid-phase object from a material in the liquid phase
US5263585	07.05.92	Microbiomed Corporation	Package for an elongated flexible fiber

US5303141	22.03.93	International Business Machines Corporation	Model generation system having closed-loop extrusion nozzle positioning
US5312224	12.03.93	International Business Machines Corporation	Conical logarithmic spiral viscosity pump
US5340433	08.06.92	Stratasys, Inc.	Modeling apparatus for three-dimensional objects
US5402351	18.01.94	International Business Machines Corporation	Model generation system having closed-loop extrusion nozzle positioning
US5418112	10.11.93	W. R. Grace & Co.-Conn.	Photosensitive compositions useful in three-dimensional part-building and having improved photospeed
US5434196	01.07.94	Asahi Denka Kogyo K.K.	Resin composition for optical molding
US5474719	14. Febr. 1991	E. I. Du Pont De Nemours And Company	Method for forming solid objects utilizing viscosity reducible compositions
US5503785	02.06.94	Stratasys, Inc.	Process of support removal for fused deposition modeling
US5587913	12.10.94	Stratasys, Inc.	Method employing sequential two-dimensional geometry for producing shells for fabrication by a rapid prototyping system
US5594652	07.06.95	Texas Instruments Incorporated	Method and apparatus for the computer-controlled manufacture of three-dimensional objects from computer data

US5690865	31.03.95	Johnson & Johnson Vision Products, Inc.	Mold material with additives
US5695707	15.05.95	3D Systems, Inc.	Thermal stereolithography
US5714541 *	16. Sept. 1996	Bayer Aktiengesells chaft	Thermoplastics having a high heat deflection temperature and improved heat stability
US5738817	8. Febr. 1996	Rutgers, The State University	Solid freeform fabrication methods
US5764521	13.11.95	Stratasys Inc.	Method and apparatus for solid prototyping
US5765740	30.06.95	Ferguson; Patrick J.	Suture-material-dispenser system for suture material
US5807437	5. Febr. 1996	Massachusett s Institute Of Technology	Three dimensional printing system
US5866058	29.05.97	Stratasys Inc.	Method for rapid prototyping of solid models
US5893404	20. Sept. 1996	Semi Solid Technologies Inc.	Method and apparatus for metal solid freeform fabrication utilizing partially solidified metal slurry
US5900207	20.05.97	Rutgers, The State University Old Queens	Solid freeform fabrication methods
US5932055	11.11.97	Rockwell Science Center Llc	Direct metal fabrication (DMF) using a carbon precursor to bind the "green form" part and catalyze a eutectic reducing element in a supersolidus liquid phase sintering (SLPS) process
US5939008	26.01.98	Stratasys, Inc.	Rapid prototyping apparatus

US5943235	27. Sept. 1996	3D Systems, Inc.	Rapid prototyping system and method with support region data processing
US5968561	26.01.98	Stratasys, Inc.	High performance rapid prototyping system
US6004124	26.01.98	Stratasys, Inc.	Thin-wall tube liquifier
US6022207	26.01.98	Stratasys, Inc.	Rapid prototyping system with filament supply spool monitoring
US6027068	19.03.98	New Millennium Products, Inc.	Dispenser for solder and other ductile strand materials
US6043322 *	23.12.97	Eastman Chemical Company	Clear polycarbonate and polyester blends
US6054077	11.01.99	Stratasys, Inc.	Velocity profiling in an extrusion apparatus
US6067480	02.04.97	Stratasys, Inc.	Method and apparatus for in-situ formation of three-dimensional solid objects by extrusion of polymeric materials
US6070107	20.05.98	Stratasys, Inc.	Water soluble rapid prototyping support and mold material
US6085957	08.04.96	Stratasys, Inc.	Volumetric feed control for flexible filament
US6095323	12.06.98	Ferguson; Patrick J.	Suture-material-dispenser system for suture material
US6119567	10.07.97	Ktm Industries, Inc.	Method and apparatus for producing a shaped article
US6127492	11.05.99	Sumitomo Chemical Company, Limited	Thermoplastic resin composition and heat-resistant tray for IC
US6129872	29.08.98	Jang; Justin	Process and apparatus for creating a colorful three-dimensional object

160 Weitere relevante Patente

US6133355	12.06.97	3D Systems, Inc.	Selective deposition modeling materials and method
US6162378	25. Febr. 1999	3D Systems, Inc.	Method and apparatus for variably controlling the temperature in a selective deposition modeling environment
US6165406	27.05.99	Nanotek Instruments, Inc.	3-D color model making apparatus and process
US6166137 *	10.12.98	General Electric Company	Poly(arylene ether)/polyetherimide blends and methods of making the same
US6175422	31.07.92	Texas Instruments Incorporated	Method and apparatus for the computer-controlled manufacture of three-dimensional objects from computer data
US6193923	14.07.99	3D Systems, Inc.	Selective deposition modeling method and apparatus for forming three-dimensional objects and supports
US6214279	02.10.99	Nanotek Instruments, Inc.	Apparatus and process for freeform fabrication of composite reinforcement preforms
US6228923	11.06.98	Stratasys, Inc.	Water soluble rapid prototyping support and mold material
US6242520 *	13.03.97	General Electric Company	Flame retardant polymer compositions with coated boron phosphate
US6252011 *	31.05.94	Eastman Chemical Company	Blends of polyetherimides with polyesters of 2,6-naphthalenedicarboxylic acid
US6257517	10.08.99	Sandvik Steel Co.	Method and apparatus for feeding welding wire

US6261077	8. Febr. 1999	3D Systems, Inc.	Rapid prototyping apparatus with enhanced thermal and/or vibrational stability for production of three dimensional objects
US6322728	09.07.99	Jeneric/Pentron, Inc.	Mass production of dental restorations by solid free-form fabrication methods
US6376571	06.03.98	Dsm N.V.	Radiation-curable composition having high cure speed
US6407163 *	07.12.99	Bayer Aktiengesellschaft	Highly impact-resistant ABS moulding materials
US6572228	31.05.01	Konica Corporation	Image forming method
US6645412	11.05.01	Stratasys, Inc.	Process of making a three-dimensional object
US6685866	09.07.01	Stratasys, Inc.	Method and apparatus for three-dimensional modeling
US6722872	23.06.00	Stratasys, Inc.	High temperature modeling apparatus
US6730252	20. Sept. 2001	Dietmar Werner Hutmacher	Methods for fabricating a filament for use in tissue engineering
US6790403	19.04.00	Stratasys, Inc.	Soluble material and process for three-dimensional modeling
US6869559	05.05.03	Stratasys, Inc.	Material and method for three-dimensional modeling
US20020013416 *	14.06.01	Noel Oscar French	High flexural modulus and/or high heat deflection temperature thermoplastic elastomers and methods for producing the same
US20020033563 *	28. Sept. 2001	Certainteed Corporation.	Apparatus for continuous forming shaped polymeric articles

US2002005 5563 *	13. Sept. 2001	Takayuki Asano	Flame retardant resin composition
US2003009 0034	01.12.00	John Hendrik	Device and method for the production of three-dimensional objects
US2004024 5663	10. Sept. 2003	Macdougald Joseph A.	Method for manufacturing dental restorations

Links und Kontakte

Linksammlung

http://www.rapidtoday.com

Eine englischsprachige Website die sich mit allem rund um Rapid Prototyping beschäftigt.

http://reprapmagazine.com/

Englischsprachige Seite, die über alles rund um das RepRap Projekt berichtet.

http://3druck.com

Deutsches 3D Druck Online Magazin

http://www.heise.de/hardware-hacks/

Bastelmagazin des heise Verlags. Absolutes Muss für Bastler, Frickler und Maker!

http://reprap.org

Open Source Projekt für einen 3D Drucker, der sich (zu Teilen) selbst herstellen kann.

http://www.fabathome.org

Ein weiteres, großes aber 2012 beendetes Open Source Projekt zum Thema 3D Druck. Trotzdem sehr lehrreich und vor allem übersichtlicher als RepRap

http://www.makexyz.com

Suchmaschine, die nach Eingabe der eigenen Postleitzahl alle registrierten 3D Drucker Dienstleister auflistet die sich in der Nähe befinden.

http://www.simple3d.com/

Eine sehr umfangreiche Seite zum Thema 3D Scan (englisch) Viele wertvolle Links und Dokumente.

Messen

Trotz Internet ist der persönliche Kontakt auf Fachmessen nach wie vor die wichtigste und beste Möglichkeit, sich einen Überblick über aktuellen Stand der Branche zu verschaffen. Nirgendwo sonst kann man in so kurzer Zeit, so viele Systeme live erleben und im Dialog neue Erkenntnisse gewinnen statt nur Daten über das Netz zu „konsumieren".

Spezialisierte Fachmessen wie die MOTEK oder die Euromold sind Goldgruben für den Informationsschürfer. Wichtiger noch: der persönliche Kontakt zu Menschen und Unternehmen, vertieft Netzwerk und erhöht die Chancen im Geschäft.

CEBIT

Nach wie vor ist die CEBIT die weltweit größte Messe für IT und hierzu zählen irgendwie auch 3D Drucker. Wer also ohnehin auf der CEBIT ist, sollte das Veranstalterverzeichnis genau studieren, viele interessante Projekte finden sich auf den Ständen von Universitäten und anderen Lehreinrichtungen und weniger bei den kommerziellen Anbietern.

Hannover Messe

Die Hannovermesse bezeichnet sich selbstbewusst als die wichtigste Industriemesse der Welt. Immerhin wurde von ihr dereinst die CeBit abgespalten, die heute allerdings kein Publikumsmagnet mehr ist. Die 3D Drucker gibt und gab es schon lange auf der Hannover Messe. Sie verbreiten sich mittlerweile sowohl auf der Hannover Messe als auch auf der CeBit.

Die Hannover Messe findet alljährlich Anfang April statt.

Link: http://www.hannovermesse.de

MOTEK Sinsheim

Die Internationale Fachmesse Motek ist weltweit die führende Veranstaltung in den Bereichen Produktions- und Montageautomatisierung, Zuführtechnik und Materialfluss, Rationalisierung durch Handhabungstechnik und Industrial Handling. Als einzigartige Branchenplattform bildet sie die ganze Welt der Automation ab. Für die Fachbesucher hat dies gegenüber den reinen Komponenten-Fachmessen oder der ausschließlichen Präsentation von speziell nach Kundenspezifikation realisierten Anlagen den Vorteil, dass Konstrukteure und Anwender hier bereichsübergreifende Lösungsansätze vermittelt bekommen, angefangen von Detaillösungen und bis hin zu schlüsselfertigen Systemlösungen.

Auf 60.000 m² Ausstellungsfläche treffen in der Messe Stuttgart Jahr für Jahr über 1.000 Aussteller auf ein internationales Publikum von rund 35.000 Fachbesuchern. Die konsequente Zielgruppenorientierung ist dabei eines der Erfolgsgeheimnisse der Motek. Schwerpunkt-Zielgruppen sind z.B. der Automobil-, Maschinen- und Gerätebau, die Elektro- und Elektronik-Industrie, die Medizintechnik und Solarproduktion, und überhaupt die metall- und kunststoffverarbeitenden Unternehmen und deren Zulieferer.

Gemeinsam mit der parallel stattfindenden Bondexpo – Internationale Fachmesse für Klebtechnologie sowie der Microsys – Technologiepark für Mikro- und Nanotechnik bildet die Motek einen schlagkräftigen Messeverbund, der den Zukunftsthemen der Branche eine ideale Plattform bietet. Ein hochkarätiges Rahmenprogramm, das sich top-aktuellen Fragestellungen widmet, vervollständigt das umfassende Angebot. (Text des Veranstalters)

Als Besucher und großes Kind, bekommt man bei dem Gang über die Motek große Augen und feuchte Hände. Das wird das Beste an moderner Feinmechanik, Handhabung und Montagetechnik aufgeboten was man in Deutschland und vermutlich weltweit zu sehen bekommen kann. Eine kleine ruhige Messe, mit reinem Fachbesuchern die es aber in sich hat. Hier findet man die Experten für eine 3D Drucker Projekt. Nichts was man hier sieht ist billig, aber alles vom Feinsten.

Link: http://www.motek-messe.de

Euromold Frankfurt

Die EuroMold hat sich seit 1994 als eine der weltweit führenden Fachmessen für Werkzeug- und Formenbau, Design und Produktentwicklung etabliert. Besonderheit der Euromold ist die Abbildung der gesamten Prozesskette "Vom Design über den Prototyp bis zur Serie". (Text des Veranstalters)

Auf der Euromold findet man etliche Händler von gebrauchten Kunststoffmaschinen jeglicher Form. So kompakt und auf enger Fläche findet man die nie wieder. Wer also mit dem Gedanken spielt sich einen Extruder oder Kunststoff verarbeitende Maschinen allgemein zu beschaffen, der liegt hier genau richtig.

http://www.euromold.com

Maker Faires
Faires (zu deutsch Messen) sprießen wie Pilze aus dem Boden, daher seine an dieser Stelle nur ein paar genannt. Welche von denen sich durchsetzen oder gar eine Leitmesse werden ist noch nicht ausgemacht. Lokale Schwerpunkte der Maker Bewegung sind jedoch das frankophone aber auch das anglophone Kanada, der Westen der USA, New York, England, die Niederlande und natürlich Deutschland. Hier eine Auswahl mehr oder weniger bekannter Events der Szene:

London Mini Maker Faire
http://makerfaireelephantandcastle.com/

World Maker Faire New York
http://makerfaire.com/
Behauptet von sich selbst, die größte Maker Messe der Welt zu sein.

Maker Faire Hannover
Hannover Congress Centrum
Glashalle & Stadtpark
Theodor-Heuss-Platz 1-3
30175 Hannover

Die Maker Faire in Hannover ist die bislang erste und wohl auch einzige Maker Messe in deutschen Raum. Durch die Nähe zur Messe Hannover, die ja auch die CEBIT und die Hannover Industrie Messe veranstaltet sowie der Sitz des heise Verlags gab es dort einen geographischen Schwerpunkt, der zur Initialzündung solcher Veranstaltungen geführt hat. Neben 3D Druckern wurden zahlreiche andere Selbstbau und Staun Projekte gezeigt, vieles hatte irgendwie mit Technik zu tun. Hannover liegt zwar in der Mitte Deutschlands, dafür aber auch von allen Ballungszentren extrem weit entfernt. Es bleibt zu hoffen dass sich andere regionale Initiativen bilden, damit mehr Menschen eine solche Veranstaltung zu geringen Reisekosten besuchen können.

Bis dahin ist die Maker Faire die bislang einzige deutsche Veranstaltung die sich explizit mit 3D Druckern beschäftigt.

Index

Stichwortverzeichnis

Weitere Titel vom Autor

Aus der Serie 3D Drucker:

Business und Technik – Professional Edition

Umfang 420 Seiten
ca. 70 Abbildungen Farbe
Ungekürzte Vollausgabe
Enthält einen Teil des Buchs „Patente & Erfindungen" und anderer
Bände vom selben Autor

Business und Technik – Student Edition

Umfang 355 Seiten
ca. 50 Abbildungen S/W
Gekürzte Ausgabe

Bauanleitung Rostock

Ausführliche deutschsprachige, bebilderte Bauanleitung für einen
Rostock Deltabot 3D Drucker . Mit Bezugsquellen und
Versandoptimierung.
Umfang ca. 50 Seiten

Deltabots

Umfang ca. 150 Seiten
Übersicht der aktuellen Modelle und Varianten der RepRap Deltabot 3D
Drucker.
Erscheinungstermin Januar 2014

Bestellmöglichkeit und mehr Informationen unter:
http://3dptb.blogspot.de

Gold aus PC und Handy

**Das erfolgreichste Buch zum Thema!
Mittlerweile in dritter überarbeiteter
Auflage!
Blick ins Buch:**

- Fast alle PC CPUs mit Goldgehalt
- Zahlreiche Abbildungen erklären genau was wie zu machen ist um Gold aus PC und Handy zu gewinnen.
- Welche Mittel und Verfahren gebraucht werden.
- Zusätzlich erklären extra auf das Buch zugeschnittene, verlinkte **Videos** den genauen Ablauf beim Goldschrott fördern.
- Bezugsquellen Scrapgold und Hilfsmittel.
- Wichtige Internetadressen
- Lieferanten für Goldschrott und Elektroschrott.
- Tabellen mit Daten zu Goldgehalt von PC Bauteile
- 185 Seiten
- 50 Abbildungen

Bestelldaten:
Titel: Gold aus PCs: Handbuch für
Einsteiger: 1
Autor: Marcel A. Buth

ISBN-10: 1478150807
ISBN-13: 978-1478150800

www.ingramcontent.com/pod-product-compliance
Lightning Source LLC
Chambersburg PA
CBHW051507170526
45166CB00001B/431